Yale Mathematical Monographs 5

James K. Whittemore Lectures in Mathematics
given at Yale University

ERGODIC THEORY, RANDOMNESS, AND DYNAMICAL SYSTEMS

by Donald S. Ornstein

New Haven and London, Yale University Press, 1974

Library of Congress catalog card number: 73-90903

International standard book number: 0-300-01745-6

Set in IBM Bold Face One type.
Printed in the United States of America by
The Murray Printing Co., Forge Village, Mass.

Published in Great Britain, Europe, and Africa by
Yale University Press, Ltd., London. Distributed in
Latin America by Kaiman & Polon, Inc., New York City;
in Australasia and Southeast Asia by John Wiley & Sons
Australasia Pty. Ltd., Sydney; in India by UBS
Publishers' Distributors Pvt., Ltd., Delhi; in
Japan by John Weatherhill, Inc., Tokyo.

CONTENTS

ACKNOWLEDGMENTS

I would like to express my appreciation and gratitude to my friends, Ya. Sinai, Nathaniel A. Friedman, Meir Smorodinsky, Paul C. Shields, Roy L. Adler, Benjamin Weiss, Yitzhak Katznelson, and Giovanni Gallavotti, who have contributed so much to the subject matter of these lectures. The isomorphism theorem for Bernoulli shifts grew out of an attempt to get a better understanding of Sinai's weak isomorphism theorem. Nathaniel Friedman worked with me to get very substantial simplifications in the original proof of the isomorphism theorem, and we jointly got a criterion to show that certain Markov shifts were Bernoullian. This paved the way for much of the later work. Meir Smorodinsky made further substantial simplifications in the isomorphism theorem and proved a strong lemma concerning entropy and ε-independence which enabled him to extend the isomorphism theorem to countably infinite partitions. I have borrowed heavily from his book Entropy and Ergodic Theory. I have also borrowed heavily from Paul Shields' book The Theory of Bernoulli Shifts.* The example of a K-automorphism that is not Bernoulli was radically simplified—the whole idea of why it worked was changed—and this came from discussions with Roy Adler and Paul Shields. The part of these lectures on K-automorphisms is largely lifted from a joint paper with Paul Shields. The part on classical systems is completely lifted from a joint paper with Benjamin Weiss and depends on work of Sinai and Anosov. Sinai's remarkable work on the hard sphere gas gives one of the nicest applications of the theory in these lectures. Gallavotti showed me how to make this application. Yitzhak Katznelson's beautiful theorem on the automorphism of the n-dimensional torus showed what could be done by combining the theory of these lectures with more classical analytic techniques. Katznelson and Weiss extended the isomorphism theorem to the case of several commuting transformations, and Gallavotti, diLiberto, and Russo applied this to the Icing model.

*This book, which gives a more detailed exposition of the underlying ideas, provides a good introduction to the subject.

PART I

The Isomorphism Theorem for Bernoulli Shifts

1. Introduction

Measure-preserving transformations

A large and central part of ergodic theory deals with abstract mathematical objects called measure-preserving transformations of a measure space. By a measure space (or probability space) we mean a set of points (like the points on the unit interval), together with a collection of subsets called measurable sets (like the Lebesgue measurable subsets of the unit interval) and a measure or probability assigned to each of these subsets (like Lebesgue measure). By a measure-preserving transformation on a measure space we mean a mapping which assigns to each point in the measure space another point in a one-to-one, onto way, and so that each measurable set is transformed onto a measurable set of the same measure.

Origins in mechanical system

Such transformations originally arose in the study of mechanical systems. For concreteness let us consider a box containing 100 hard spheres bouncing around. Each configuration is determined by the position and velocity of each sphere and hence by 600 numbers. These configurations (which could be thought of as points in a 600 dimensional Euclidean space) form the points of our measure space (sometimes called the phase space of the mechanical system).

For each measurable subset of phase space, we can consider the probability that if we look at our system its configuration will belong to that set, and this gives us our measure (in this case it is the same as Euclidean measure). If the system has a certain configuration, then the laws of mechanics determine its configuration one unit of time later. This is our transformation. This point of view goes back to Gibbs.

"Isomorphism"

Since ergodic theory was an attempt to study the statistical properties of mechanical systems, we ignore events that happen with probability 0 and we consider only the statistical structure of the transformation (ignoring its topological structure, etc.). Hence we say that two measure-preserving transformations are statistically the same or "isomorphic" if we can find a one-to-one correspondence between all (but a set of measure 0) of the points in each measure space (or phase space), so that corresponding sets have the same measure (or probability), and the corresponding points are transformed in the same way (that is into corresponding points).

Origins in random processes

Measure-preserving transformations also arise in the study of stationary random processes. A stationary random process can be thought of as a box that prints out one letter each unit of time, where the probability of printing out a given letter may depend on the letters already printed out, but is independent of the time (that is, the mechanism in the box does not change).

Example 1: The box contains a roulette wheel. We spin the wheel once each unit of time and print out the result. (We call such a process an independent process.)

Example 2: The box contains a roulette wheel. We look at all possibilities for three consecutive spins of the wheel, and divide these into two classes. Each time we spin the wheel, we look at the last three spins and print out 1 if they fall in the first class, and 2 if they fall in the second class. (We call this process a finite coding of an independent process.)

Example 3: The box contains two coins, one of which is biased so that the probability of heads is not $\frac{1}{2}$. We divide all sequences of heads and tails of length three into two classes. At each unit of time, we look at the sequence of heads and tails which the box has printed out in the last three times. If the sequence lies in the first class, we flip the first coin and print out heads; if it comes up heads and tails, it comes up tails. If the previous three print-outs were in the second class, then we would use the second coin. This is an example of a three-step Markov process. That is, the last three print-outs determine the probability of printing out "heads," but if we know the last three print-outs, the conditional probability of "heads" is unaffected by an additional knowledge of what the process printed out in the past. (Note that example 2 need not be an n-step Markov process for any n.)

Example 4: The box contains a mechanical system such as a gas. We make a fixed measurement on the system, at each unit of time, which has only a finite number of possible outcomes, and print out the outcome of the measurement. (If the outcome of the measurement were real valued, we could divide the line into a finite number of sets, and print out which set the measurement fell into.)

Example 5: A teleprinter—this prints out letters where the probability of printing out a given letter depends on what has already been printed. (Many possibilities will have probability 0 because they will not make sense.)

The mathematical model for a stationary random process is a measure-preserving transformation of a measure space and a partition of the measure space into a finite number of disjoint sets.

We can see this as follows: If we let the process run forever (with no beginning and no end) we get an infinite sequence of outputs. These sequences will be the points of our measure space. The probabilities (or measures) of those sets of sequences, formed by specifying the print-out at some finite number of specified times, is determined by (and determines) the statistics of the process and determines the measure on our measure space. The transformation corresponds to the progression of time, and thus shifts each sequence. The partition is the partition according to the output at time 0. If we are given only the transformation and the partition, then for each point in our measure space we can recover the sequence of outputs to which it corresponded by seeing which element of the partition the point is in at various times. (That is, we apply different powers of the transformation to our point and see what element of the partition the resulting point lies in.)

This point of view goes back to Kolmogorov and is especially natural if our process is a measurement on a mechanical system (with a finite number of possible outcomes). Then our measure space and transformation are already given. Our partition is the one formed by putting all those configurations for which the measurement has the same outcome into one element.

Origins in abstract mathematics

Any automorphism of a compact group is a measure-preserving transformation (with respect to Haar measure).

Measure-preserving transformations also arise in number theory, for example in the continued fraction expansions of numbers.

Abstract ergodic theory, Bernoulli shifts

From the abstract point of view there is a class of transformations, the so-called Bernoulli shifts, which play a central role. The Bernoulli shifts are those transformations that arise from independent processes (see example 1). More specifically: Start with a "roulette wheel" with k slots and call the fraction of the circumference occupied by the ith slot p_i. This gives us an independent process, and from this process we get a measure-preserving transformation. Thus, if we start with k positive numbers $p_1 \ldots p_k$ that add up to 1, we construct a transformation called the Bernoulli shift, $(p_1 \ldots p_k)$ or $B_{(p_1 \ldots p_k)}$.

Bernoulli shifts play a central role in ergodic theory for much the same reason that independent processes play a central role in probability theory —they are in some sense the simplest transformations, and one feels that they must be understood before we can understand the more complicated ones.

A very easy mathematical result is that the Bernoulli shift $(p_1 \ldots p_k)$ is isomorphic to the following transformation of the unit square: We divide the square into k rectangles where the ith rectangle has a base of width p_i

and height 1. We squeeze the height of the ith rectangle and expand its width so that it now has height p_i and width 1. We now reassemble these rectangles to form the square again by putting the first on the bottom, the second on top of it, etc.

We will draw the picture for $B_{\left(\frac{1}{2},\frac{1}{2}\right)}$:

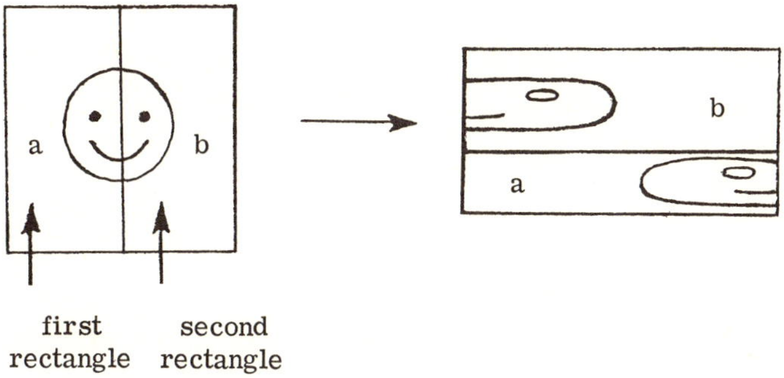

The isomorphism problem

When Halmos wrote his book on ergodic theory [1], one of the main problems was the following: Are all Bernoulli shifts isomorphic? In particular, is the Bernoulli shift $\left(\frac{1}{2},\frac{1}{2}\right)$ isomorphic to the Bernoulli shift $\left(\frac{1}{3},\frac{1}{3},\frac{1}{3}\right)$? Up to this point there were various properties of transformations that were known, and because of these a large number of transformations could be distinguished from one another (by properties such as ergodicity, mixing, weak mixing). None of these properties, however, distinguished any two Bernoulli shifts (two Bernoulli shifts even induced isomorphic unitary operators).

The breakthrough in this area came when Kolmogorov [36, 37] introduced a new invariant called entropy, motivated by Shannon's work on information theory. This invariant was easy to compute for the Bernoulli shift $(p_1, \ldots, p_k)$ and is simply $-\sum_{i=1}^{k} p_i \log p_i$. (Thus the Bernoulli shift $\left(\frac{1}{2}, \frac{1}{2}\right)$ is not isomorphic to the Bernoulli shift $\left(\frac{1}{3}, \frac{1}{3}, \frac{1}{3}\right)$.)

There are still an infinite number of Bernoulli shifts with the same entropy and the question remained: Are all of these the same? Or are they all different? Was there some yet to be discovered property that would distinguish more Bernoulli shifts?

There were many beautiful and deep results along these lines by Meshalkin [39], Sinai [38], and Adler and Weiss [45]. Our first aim in these lectures [7] will be to show:

THEOREM 1. Two Bernoulli shifts with the same entropy are isomorphic.

Thus, to see if two Bernoulli shifts are isomorphic it is enough to calculate $p_1 \log p_1 + p_2 \log p_2 + \ldots + p_k \log p_k$.

Abstract ergodic theory, Kolmogorov transformations

There is another class of transformations, introduced by Kolmogorov and now called K-transformations, that also plays a central role.

The K-transformations are those which arise from processes satisfying the so-called 0 - 1 law of probability theory. [The 0 - 1 law says that if an event has the property that knowledge of what the process did in the arbitrarily distant past can be used to predict it (or shed any light on whether or not it will occur), then the event has probability 0 or 1. These processes are in some sense completely nondeterministic.]

The K-transformations play a central role because of the large number of examples that can be shown to be K. Bernoulli shifts are K. Ergodic automorphisms of compact processes are K. Sinai was able to show that the transformation arising from the hard sphere gas in a rectangular box is K [32, 33].

Kolmogorov conjectured that entropy classified the K-transformations, and hence that K-transformations were all Bernoulli. In Part III we will show [12]:

THEOREM 2. There is a K-transformation that is not a Bernoulli shift.

It was once thought that K-transformations were the most random possible. The above theorem shows this to be false.

Behind Theorem 2 lies a general method for constructing certain counterexamples. The method has so far been used to get K-transformation with no square root [13, 54], ones not isomorphic to their inverses (Shields and myself) [16], and a counterexample to Pinsker's conjecture [14].

Abstract ergodic theory, Bernoulli flows

A mechanical system actually evolves continuously with time, that is, for each time t we have a transformation S_t which tells us how each configuration will evolve in time t. We call such a family of transformations, S_t, a flow.

An extension of the methods used in proving Theorem 1 can be used to show that [10, 22].

THEOREM 3.* There is a flow S_t, such that S_1 is a Bernoulli shift (call such a flow a Bernoulli flow).

Thus, for the first time it became theoretically possible at least that Bernoulli shifts could arise in the study of mechanical systems.

A further extension of the method used in proving Theorem 1 yields [11]:

THEOREM 4. There is only one Bernoulli flow (of finite entropy). That is (except for a trivial change in the time scale), any two Bernoulli flows (of finite entropy) are isomorphic. Similarly, there is only one Bernoulli flow of infinite entropy.

The Bernoulli flow in Theorem 3 has a very simple description, and hence all Bernoulli flows have a simple description.

---

*One consequence of Theorem 3 is that the Bernoulli shift $(\frac{1}{2},\frac{1}{2})$ has a square root. This fact is also an immediate consequence of Theorem 1, but it was not known before Theorem 1 was proved.

A criterion for showing transformations to be Bernoulli

An analysis of the ideas involved in proving Theorem 1 yields a criterion for showing that certain transformations are Bernoulli [15, 11]. This criterion provides the link between the abstract theory and its applications. Because of it, certain transformations which were previously studied quite apart from Bernoulli shifts, turn out quite surprisingly to be Bernoulli shifts. Work in this direction is just beginning, and I think that there is room here for a great deal of further development.

I will now list some transformations that are now known to be Bernoulli:

(1) the transformation arising from a mixing multi-step Markov process (see example 3) (this result is due to N. Friedman and myself [15]),

(2) subshifts of Bernoulli shifts [9],

(3) ergodic automorphisms of the n dimensional torus (the criterion applies easily for dimension 2 or 3. For higher dimensions the theorem, in its full generality, is due to Katznelson [50]),

(4) the continued fraction transformation (this result is due to several people [28, 29, 30, 31]),

(5) geodesic flow on a surface of negative curvature is a Bernoulli flow (this result is due to B. Weiss and myself [17]),

(6) the flow arising from a hard sphere gas in a rectangular box (the energy of the system is fixed). (This result is due to Sinai, Gallavotti, and myself.)

I think that the last two results are important because of the promise they hold for generalizations and extensions to other mechanical systems. (Geodesic flow on a surface of negative curvature is some sort of mechanical system, that is, it represents the motion of a particle constrained to move on a surface and is a mathematically simple version of the hard sphere gas.)

Geodesic flow on surfaces of negative curvature was studied extensively by Hopf, Hedlund, Gelfond and Fomin, Sinai, and Anosov, and various properties of the flow (ergodicity, mixing, Lebesgue spectrum, K) were established [46, 47, 48, 56, 57, 58]. Sinai showed that the hard sphere gas was K [32, 33].

The significance of (5) is that we are now able to determine exactly what the flow is (up to isomorphism). We thus know exactly—in some sense—the probabilistic structure of the flow.

Random processes again

The methods arising from the proof of Theorem 1 can be used to get some sort of overall view of the class of stationary random processes, and more insight into the possible kinds of non-determinancy.

We first need the following notion of distance between processes (I will call this the $\bar{d}$ distance): We say that two processes are close if one process is the same as the other, modified very infrequently.

For example, if we make a measurement on a mechanical system, then because of experimental error the process we actually get and the one we should get are close in the $\bar{d}$ sense, but not identical.

The process arising from flipping a coin where the front and back are different is only close in the $\bar{d}$ sense to the process obtained by flipping an absolutely symmetric coin.

With the above notion of distance we get the following overall view of processes: There is a class of processes called B-processes. These are defined to be those processes whose model is a Bernoulli shift and any arbitrary partition.

THEOREM (A). Any process that can be approximated arbitrarily well by B-processes is a B-process.

THEOREM (B). The B-processes are exactly those processes that can be approximated arbitrarily well by finite codings of a roulette wheel (see example 2).

THEOREM (C). The B-processes are exactly those processes that can be approximated arbitrarily well by multi-step mixing Markov processes (see example 3).

Most of the processes one encounters in practice are B-processes. The B-processes are the most random possible. Theorem (C) implies that they are the only processes that can be approximated arbitrarily well by processes having a finite memory.

Any measurement made on a hard sphere gas or on a geodesic flow on a surface of negative curvature is a B-process (because of (5)). Thus, even though the flow evolves in a completely deterministic way, as soon as we make a gross measurement (one with only a finite number of possible outcomes, e.g., look at the system on a TV screen) the resulting process is the most random possible, and is essentially indistinguishable from a finite coding of a roulette wheel or a multi-step Markov process. This helps explain why statistical methods can play such an important role in the study of mechanical systems.

We next come to the class of K-processes, which we have already discussed (these include the B-processes). Our counterexample to Kolmogorov's conjecture shows that there is a K-process that is not a B-process. Thus, there is a process satisfying the 0 - 1 law (i.e., completely nondeterministic) that cannot be approximated by multi-step Markov processes or by finite codings of roulette wheels. It is thus false that all completely nondeterministic processes could be regarded as having as their random mechanism a roulette wheel.

Lastly, we come to the processes of 0 entropy. These are deterministic in the sense that we can predict the future with probability 1 by knowing the past.

It was conjectured by Pinsker that every process was made up in a simple way from a process of 0-entropy, and a K-process was shown to be false using the same kind of construction as that in Theorem 2 [13, 14].

Codes

It should be mentioned that to say that two processes yield isomorphic transformations is the same as saying that one process can be coded into the other by an invertible code (which can be approximated arbitrarily well by finite codes. See example 2). A more precise statement is the following:

For the sake of notation, assume that our first process prints out only 0,

1's and the second 1, 2, 3's. A code of length n is a prescription for associating to each sequence of 0, 1's a sequence of 1, 2, 3's as follows: Divide all finite sequences of 0, 1's of length 2n + 1 into 3 classes. Look at the 2n + 1 consecutive terms in our 0, 1 sequence whose middle term has index i, and let the ith term in our 1, 2, 3 sequence be 1, 2, or 3 according to which class the above 2n + 1 consecutive 0, 1's belong (such a coding can be physically realized if we allow a time delay and give another process).

We say that a sequence of codes of length n converge on our first process if for each sequence of 0, 1 (except for a collection of measure 0, under the first process) and each i, all but a finite number of codes will code the above sequence of 0, 1's into a sequence of 1, 2, 3's with the same i^{th} term. This implies that any 2 sufficiently long codes in our sequence agree with high probability. We call the limit of a convergent sequence of codes an infinite code.

If two processes yield isomorphic transformations, then the second can be obtained by an infinite coding of the first, and the first can be recaptured by then applying an infinite code to the second. The B-processes are exactly those processes that can be obtained as an infinite coding of an independent process (i.e., some roulette wheel).

Other transformation groups

So far, we have always interpreted our transformation as representing the passage of time. Because of recent work on infinite partical systems by Spitzer, Rouelle, Debrushkin, and others, it seems that it would be useful to study the spacial shift transformations. We are thus led to the study of several commuting transformations (the shifts in the x, y, and z directions).

Recently, Y. Katznelson and B. Weiss [23] showed that Theorem 1 holds for several commuting transformations. I expect this line of research to prove very fruitful. Gallavotti, diLiberto, and Russo applied these results to the Icing model, showing certain states to be Bernoullian.

Mechanical systems can have more complicated symmetry groups, and in this direction Krieger extended the isomorphism theorem to certain other groups.

An outline of the book

We will first trace a minimal path to the isomorphism theorem for Bernoulli flows of finite entropy. This takes up a large part of the book. We first introduce (sec. 3) the $\bar{d}$ distance between sequences of partitions. We then introduce (sec. 3) the notion of a process (a transformation and a partition) being "finitely determined" or F.D. We then show (sec. 4) that independent processes are F.D. This is the only property of independence used to prove the isomorphism theorem for Bernoulli shifts (sec. 4), and we therefore get this theorem as a corollary of Theorem 12 (sec. 4), which says that two transformations of the same entropy having F.D. generators are isomorphic. We next show (sec. 5) that if T is a Bernoulli shift and P any finite partition, then the process P,T is F.D. (Hence, factors of Bernoulli shifts are Bernoulli.) We then show that if a process satisfies a condition called "very weak Bernoulli" or V.W.B., then it is F.D. (sec. 7).* We then examine a specific flow and show that it is a Bernoulli flow by exhibiting a V.W.B. generator (sec. 10). We next use the fact that any partition under a Bernoulli shift is F.D., and imitate the proof of the isomorphism theorem for transformations with F.D. generators to get the isomorphism theorem for Bernoulli flows of finite entropy (sec. 11). (We cannot work with independent generators because even if P is independent under S_1, it will not be independent for S_t for small t.)

In sec. 9 we extend the isomorphism theorem for Bernoulli shifts of finite entropy to Bernoulli shifts of infinite entropy. In sec. 12 we parallel this extension for flows.

In sec. 6 we show that the B-processes are closed in the $\bar{d}$-metric. (This comes easily from the main lemmas of the isomorphism theorem for Bernoulli shifts.) In sec. 8 we apply the criterion of V.W.B. to show that multi-step mixing Markov processes are B-processes. We then (sec. 8)

---

*It was proved recently by Weiss and Ornstein [62] that if P,T is F.D., then P,T is V.W.B.

apply the above results and the fact that B-processes are F.D. to show that the B-processes are exactly the closure in the $\bar{d}$-metric of the multi-step mixing Markov processes.

In Part II we give the example of a K-automorphism that is not Bernoulli. In Part III we use the V.W.B. criterion to show that the ergodic automorphisms of the 2-dimensional torus and the geodesic flow on a surface of negative curvature are Bernoulli.

2. Notation

All our transformations will be 1 - 1, invertible and measure-preserving, and will act on a measure space $X = (X,\sigma,m)$ such that m(X) 1. We will further assume, without explicitly stating so, that our transformations are ergodic Lebesgue spaces as defined in Appendix A. σ will be some complete σ-algebra, of Lebesgue measurable sets, but not necessarily all Lebesgue measurable sets.

A partition P of X will mean an <u>ordered finite partition</u> whose atoms are $P^1 \ldots P^k$. We will denote the number of atoms in P by $|P|$.

The distribution of P or dist P will be the vector $m(P^1) \ldots m(P^k)$. If E is a set, dist(P/E) will denote $\frac{m(P^1 \cap E)}{m(E)}, \ldots, \frac{m(P^k \cap E)}{m(E)}$.

If $|P| = |Q| = k$, then $|\text{dist } P - \text{dist } Q|$ will denote

$\sum_{i=1}^{k} |m(P^i) - m(Q^i)|$, and $|\text{dist }(P/E) - \text{dist }(Q/F)|$ will denote

$$\sum_{i=1}^{k} \left| \frac{m(P^i \cap E)}{m(E)} - \frac{m(Q^i \cap F)}{m(F)} \right| .$$

If P and Q are defined on the same space X, then $|P - Q|$ will denote the measure of the points x in X such that $x \in P^i$, $x \in Q^j$, and $i \neq j$.

[(Thus $|P - Q| = m(X - \bigcup_{i=1}^{k} P^i \cap Q^i)$)].

If $\{\alpha_i\}_0^{n-1}$ is a sequence, then we define the distribution of sequences of length u in $\{\alpha_i\}_0^{n-1}$ to be the probability measure on sequences of length u obtained by giving $\{\beta_i\}_0^{u-1}$ the measure r/n−u where r is the number of j, such that $\alpha_{j+i} = \beta_i$, i = 0, . . . , u − 1.

$P \vee Q$ will denote the ordered partition whose atoms are $P^i \cap Q^j$ (where the ordering is lexicographical ordering of the pairs i,j). Similarly, $\bigvee_0^n T^i P$ (or $\bigvee_0^n P_i$) denotes an ordered partition (the P_i are partitions), $\bigvee_{-\infty}^{\infty} T^i P$ denotes the smallest σ-algebra containing all of the $T^i P$ (and contains all sets of measure 0).

The <u>P-name of x</u> will denote the sequence $\{\alpha_i\}_{-\infty}^{\infty}$; where $\alpha_i = j$ if $T^i x \in P^j$. The <u>P-n-name of x</u> will be $\{\alpha_i\}_0^{n-1}$, <u>the P-name of an atom,</u> $A \in \bigvee_0^n T^{-i} P$, will be the P-n-name of the $x \in A$ ($\bigvee_0^n T^{-i} P$ is the partition of X into P-n-names).

$P,T \sim R,S$ will mean that dist $\bigvee_0^n T^i P$ = dist $\bigvee_0^n S^i R$ for all n. (Thus, P,T and R,S represent the same process.)

If $\mathscr{C}$ is a collection of sets we will call the measure of the union of the sets in it $m(\mathscr{C})$, or the measure of $\mathscr{C}$.

If $\mathscr{A}$ is a sub-σ-algebra of (X, σ, m), and P a partition of (X, σ, m), we will say that $\mathscr{A} \supset^{\varepsilon} P$ if there is a $\hat{P}$, measurable with respect to $\mathscr{A}$, and $|\hat{P} - P| < \varepsilon$. If P is measurable with respect to $\mathscr{A}$, we will write $\mathscr{A} \supset P$.

We will sometimes write $\bigvee_{-K}^{K} T^i Q \supset^{\varepsilon} P$. This will make sense if we regard $\bigvee_{-K}^{K} T^i Q$ as denoting the algebra of sets generated by the partition $\bigvee_{-K}^{K} T^i Q$.

The letters P, Q, and R will always denote partitions. The letters S, T, and U will denote transformations.

3. Definitions

<u>$\bar{d}$ distance between processes.</u> In proving the isomorphism theorem we will need the following notion of distance between sequences of partitions or distance between processes:

Let P_i, $i \leq i \leq n$ be a sequence of ordered partitions of X, each having k atoms. Let Q_i, $1 \leq i \leq n$ be a sequence of ordered partitions of Y, each having k atoms. We define $\bar{d}(\{P_i\}_1^n, \{Q_i\}_1^n)$ as follows: Let φ and ψ be 1 – 1 measure-preserving maps of X and Y respectively onto Z. Let $\bar{d}_{\varphi,\psi}(\{P_i\}_1^n, \{Q_i\}_1^n) = \frac{1}{n}\sum_{i=1}^{n} |\varphi(P_i) - \psi(Q_i)|$ (see "Notation" for $|P - Q|$). We define $\bar{d}(\{P_i\}_1^n, \{Q_i\}_1^n$ as $\inf_{\varphi,\psi} \bar{d}_{\varphi,\psi}(\{P_i\}_1^n, \{Q_i\}_1^n)$.

We now define $\bar{d}((P,T), (Q,S))$ as $\sup_n \bar{d}(\{T^iP\}_1^n, \{S^iQ\}_1^n)$.

Actually it is not hard to see that the above sequence has a limit which is equal to the sup, because if $\bar{d}(\{T^iP\}_1^r, \{S^iQ\}_1^r) = \alpha$, then for all s, $\bar{d}(\{T^iP\}_1^{sr}, \{S^iQ\}_1^{sr}) \geq \alpha$, and hence for n large enough $\bar{d}(\{T^iP\}_1^n, \{S^iQ\}_1^n) \geq \alpha - \varepsilon$. It is the existence of the above limit which shows that the above definition of $\bar{d}$ satisfies the triangle inequality and hence yields a metric.

The above definition of $\bar{d}$ is all that we will use in our proofs. It might be helpful to point out that $\bar{d}$ gives us a very reasonable notion of distance between processes. Roughly speaking, if we alter the output of a process very infrequently, then we do not change the process very much in the $\bar{d}$ sense.

There are two easy results (whose proofs will be left to an appendix) which show that the $\bar{d}$ distance between processes has some physical meaning:

(1) $\bar{d}((P,T), (Q,S))$ is the inf of the α such that we can find an ergodic U acting on Z and partitions $\overline{P}$ and $\overline{Q}$ of Z such that (a) $|\overline{P} - \overline{Q}| = \alpha$, and (b) $\overline{P},U \sim Q,T$ and $\overline{Q},U \sim Q,S$.

(We thus have one process U that prints out 2 letters each unit of time. If we look only at the first letter we get the P,T process, and if we look only at the second letter we get the Q,S process. Furthermore, the 2 letters will be the same with probability $1 - \alpha$.)

(2) $\bar{d}((P,T), (Q,S))$ is the sup of the α satisfying: given ε there is an N such that if $K > N$, then we can find 2 collections C_1 and C_2 of sequences

of length K, such that K consecutive outputs of P,T will lie in C_1 with probability $> 1 - \varepsilon$, and K consecutive outputs of Q,S will lie in C_2 with probability $> 1 - \varepsilon$, and any sequence in C_1 differs from any sequence in C_2 in more than αK places.

(The reason for introducing the second definition of $\bar{d}$ is the following: It is clear from the first definition that if the $\bar{d}$ distance is small the two processes are in some sense close to each other. However, it is not clear what it means for $\bar{d}$ to be large. The second definition says that if $\bar{d} > \alpha$ then you could tell the two processes apart even if each of the processes were changed $< \frac{1}{2}\alpha$ percentage of the time.)

We will now say what the $\bar{d}$ distance between two processes means in terms of a single infinite print-out of each process. This can be done because the Birkhoff ergodic theorem tells us that with probability 1 an infinite print-out of an ergodic process tells us everything about the process. Let us state this more precisely. We will call an infinite sequence $\{\alpha_i\}$ representative of the process P,T if each finite sequence $\{\beta_i\}_1^K$ occurs in $\{\alpha_i\}$ with a frequency equal to the measure of $\bigcap_1^K T^i P^{\beta_i}$ (P^{β_i} being the β_i atom of P). The Birkhoff ergodic theorem says that if T is ergodic then for a. e. x, the P-name of x is representative of P,T.

With this in mind it is not hard to see that if $\bar{d}((P,T), (Q,S)) = \alpha$, then for a. e. x, the P-name of x (that is: almost every print-out of the process) can be altered on a set of density $< \alpha + \varepsilon$ for every $\varepsilon > 0$ to obtain a sequence which is representative of (Q,S). Furthermore, if there are representative sequences of (P,T) and (Q,S) that differ on a subsequence of density $< \beta$, then $\bar{d}((P,T), (Q,S)) < \beta$.

This follows easily from (1) and (2) above, and since this fact is never again used we omit its proof.

The "vague" topology. There is another topology on processes called the "vague" topology. Let P_i, T_i and P,T be ergodic processes, and assume that all of the P_i and P are ordered partitions with the same number of atoms. We say that P_i, T_i converges to P,T in the vague topology if for each fixed K, the measure $u_{i,K}$ that P_i, T_i put on sequences of length K converges to u_K, the measure that P,T puts on sequences of length K. (Note that convergence in $\bar{d}$ implies convergence in the vague topology, but not conversely.)

"Finitely determined processes." We will say that P,T is finitely determined or F.D. if given ε there is a δ and an n such that if $\bar{P},\bar{T}$ satisfies (a) $|\bar{P}| = |P|$, (b) $|E(\bar{P},\bar{T}) - E(P,T)| < \delta$, and (c)

$$\left|\text{dist} \bigvee_0^n T^i P - \text{dist} \bigvee_0^n \bar{T}^i \bar{P}\right| < \delta \text{, then } \bar{d}((P,T),(\bar{P},\bar{T})) < \varepsilon.$$

Remark. We will show that if $T^i P$ is independent, then P,T is F.D. This is the only property of independence that we use in the isomorphism theorem. Hence, the isomorphism holds transformation with a F.D. generator. This will give us the extensions of the isomorphism theorem. We will show that any partition of a Bernoulli shift is F.D., and hence factors of Bernoulli shifts are Bernoulli. We will also show that another condition called "very weak Bernoulli" implies F.D. We then find a "very weak Bernoulli" generator for certain flows, automorphisms of the torus, etc.

Remark. Condition (c) in the definition of F.D. could be replaced by $\bar{d}(\{T^i P\}_1^n, \{\bar{T}^i \bar{P}\}_1^n < \delta$. This would make the definition more symmetric (i.e., if $|P| = |\bar{P}|$, and if $\bar{P},\bar{T}$ imitates P,T, in the $\bar{d}$ sense up to time n, and has good enough entropy, then it will imitate P,T forever in the $\bar{d}$ sense).

ε-independence. We say that P is ε independent of Q or $P \perp^{\varepsilon} Q$ if there is a collection C of atoms Q such that $m(C) > 1 - \varepsilon$ and if $A \in C$ then $|\text{dist } P/A - \text{dist } P| < \varepsilon$. (Note that P and Q are independent if and only if $P \perp^0 Q$.)

We say $T^i P$ is ε independent if $T^n P \perp^{\varepsilon} \bigvee_0^{n-1} T^i P$ for all n.

Remark. There is a more symmetric definition of ε-independence (which we will not use). We say $P \overset{\varepsilon}{\perp} Q$ if $\sum_j \sum_i |m(P_i \cap Q_j) - m(P_i)m(Q_j)| < \varepsilon$. (It is easy to see that $P \overset{\varepsilon}{\perp} Q \rightarrow P \overset{2\varepsilon}{\perp} Q$ and $P \overset{\varepsilon}{\perp} Q \rightarrow P \overset{\sqrt{\varepsilon}}{\perp} Q$.)

Very Weak Bernoulli or V.W.B. P,T is V.W.B. if given ε, then (A): there is an n such that for all m we can find a collection C_m of atoms of $\bigvee_{-m}^{-0} T^i P$ such that $m(C_m) > 1 - \varepsilon$, and if $A \in C_m$, then $\bar{d}(\{T^i P\}_1^n, \{T^i P/A\}_1^n) < \varepsilon$.

4. Proof of the isomorphism theorem

Part I.

Let us recall that $P \overset{0}{\perp} Q \leftrightarrow E(P \vee Q) - E(Q) = E(P)$. By using the idea of ε-independence we can get an extension of this result.

LEMMA 1. Given $\varepsilon > 0$ and an integer k, there is a $\delta(\varepsilon,k) > 0$ such that if P has k atoms and if $|(E(P \vee Q) - E(Q) - E(P)| < \delta(\varepsilon,k)$, then $P \overset{\varepsilon}{\perp} Q$.

Remark. The other implication extends in a similar way, but we will not need this.

Proof. (1) If P is not ε independent of Q, then there is an atom of P (say P^1) and a collection C of atoms of Q such that $m(C) > \varepsilon/k$, and if Q^i is in C, then $\left| m(P^1 \cap Q^i)/m(Q^i) - m(P^1) \right| > \varepsilon/k$.

(2) Let $\overline{Q}$ be the partition with three atoms; the complement of union of the Q^i in C, the Q^i in C for which $m(P^1 \cap Q^i)/m(Q^i) - m(P^1)$ is > 0, and those Q^i for which $m(P^1 \cap Q^i)/m(Q^i) - m(P^1) < 0$. It is easy to see that P is not $\varepsilon/2k$ independent of $\overline{Q}$.

(3) $E(P \vee \overline{Q}) - E(\overline{Q})$ is a function of 3k variables (the measures of the atoms of $P \vee \overline{Q}$). The set $F(\varepsilon k)$ in 3k Euclidean space such that P is not $\varepsilon/2k$ independent of $\overline{Q}$ is compact. Furthermore, $|E(P \vee \overline{Q}) - E(\overline{Q}) - E(P)| \neq 0$ on $F(\varepsilon,k)$. Therefore, there is a $\delta(\varepsilon,k) > 0$ such that $|E(P \vee \overline{Q}) - E(\overline{Q}) - E(P)| > \delta(\varepsilon,k)$ on $F(\varepsilon,k)$.

(4) By using result (5) of Appendix 2 combined with (3) above, we get that if P is not ε independent of Q, then $|[E(P \vee Q) - E(Q)] - E(P)| > \delta(\varepsilon,k)$.

Remark. Smorodinsky has shown that $\delta(\varepsilon,k)$ can be chosen independently of k [18].

LEMMA 2. If T^iP are independent, then given ε there is a δ such that if (a) $|Q| = |P|$, and if (b) $|E(Q,S) - E(P,T)| < \delta$, and (c) $|\text{dist } P - \text{dist } Q| < \delta$, then $\bar{d}((P,T),(Q,S)) < \varepsilon$.

Proof. Let $|P| = k$. (1) Using our previous lemma, we get that if (a) $|E(Q,S) - E(Q)| < \delta(\varepsilon,k)$, then (b) $S^nQ \perp^{\varepsilon} \bigvee_{j=0}^{n-1} S^jQ$ for all n.

(2) We will finish the proof of the lemma by proving the following statement: If $|\text{dist } P - \text{dist } Q| < \varepsilon$, and if T^iP are independent, and if $S^nQ \perp^{\varepsilon} \bigvee_0^{n-1} S^jQ$, then $\bar{d}((Q,S),(P,T)) < 3\varepsilon$.

To prove this assertion we will, for each n, find 2 sequences of partitions, $\{Q_i\}_1^n$ and $\{P_i\}_1^n$, of the same space such that $\text{dist} \bigvee_1^n Q_i = \text{dist} \bigvee_1^n S^iQ$, and $\text{dist} \bigvee_1^n P_i = \text{dist} \bigvee_1^n T^iP$, and $|P_i - Q_i| < 3\varepsilon$.

We will prove this last statement by induction. Suppose we have already picked P_i and Q_i for $1 \leq i \leq n-1$. We will pick P_n and Q_n as follows: We will pick P_n so that the distribution of P_n on each atom of $\bigvee_1^{n-1} (Q_i \vee P_i)$ is the same as the distribution of P. (This will insure that $\text{dist} \bigvee_1^n P_i = \text{dist} \bigvee_1^n T^iP$ because T^iP are independent.) We will pick Q_n so that if A is an atom of $\bigvee_1^{n-1} (P_i \vee Q_i)$, and $\bar{A}$ the atom of $\bigvee_1^{n-1} Q_i$ that contains A, and $\tilde{A}$ the atom of $\bigvee_1^{n-1} S^iQ$ corresponding to $\bar{A}$, then $\text{dist}(Q_n/A) = \text{dist}(S^nQ/\tilde{A})$. (This will insure that $\text{dist} \bigvee_0^n Q_i = \text{dist} \bigvee_0^n S^i Q$.) Because $S^n Q \perp^{\varepsilon} \bigvee_1^{n-1} S^i Q$ we have that there is a collection C of atoms of $\bigvee_1^{n-1} S^i Q$, such that $m(C) < \varepsilon$ and if $\tilde{A}$ is not in C then $|\text{dist } S^n Q/\tilde{A} - \text{dist } Q| < \varepsilon$. Thus, if Q_n is chosen as above, we have that there is a collection C_1 of atoms of $\bigvee_1^{n-1} (P_i \vee Q_i)$ such that $m(C_1) < \varepsilon$, and if A is not in C_1, then $|\text{dist } Q_n/A - \text{dist } Q| < \varepsilon$. Because $|\text{dist } P - \text{dist } Q| < \varepsilon$, we get that if A is not in C_1, then $|\text{dist } (Q_n/A) - \text{dist } P| < 2\varepsilon$. Hence if A is not in C_1 we can choose P_n and Q_n so that on A, P_n and Q_n differ on a set of measure $< 2\varepsilon\, m(A)$. This implies that P_n and Q_n can be chosen with the additional property that $|P_n - Q_n| < 3\varepsilon$.

COROLLARY 3. If $T^i P$ are independent, then P,T is F.D.

Part II.

The following lemma is put in for expository purposes. What we really need is a stronger form of this lemma (Lemma 7), which will be proved later (independently of this lemma).

LEMMA 4. Let R,S be F.D., and $E(R,S) \leq E(T)$. Then, given $\delta > 0$, we can find P' such that (1) $\bar{d}[(R,S),(P',T)] < \delta$.

Proof of Lemma 4. Because S is F.D., it will be enough to show that, given $\bar{\delta}$ and u, we can find P' satisfying (1) $\left|\text{dist } (\bigvee_0^u T^i P') - \text{dist } (\bigvee_0^u S^i R)\right| < \bar{\delta}$, and (2) $|E(P',T) - E(R,S)| < \bar{\delta}$. (This is the only place where we use the fact that (S,R) is F.D.)

Choose a Q such that $E(R,S) - E(Q,T) = \beta > 0$ where $\beta < \frac{\bar{\delta}}{100}$.

Choose $\gamma < \bar{\delta}$ and such that $|Q' - Q| < \gamma$ implies that $E(Q',T) \geq E(Q,T) - \frac{\bar{\delta}}{100}$. (We will prove (2) by showing $\bigvee_{-\infty}^{\infty} T^i P' \supset {}^{\gamma}Q$).

We will call an atom in $\bigvee_0^n T^{-i} Q$ good (or good Q-n-atoms) if its measure is between $(\frac{1}{2})^{[E(Q,T) \pm \beta/100]n}$.

We will call an atom in $\bigvee_0^n S^{-i} R$ good (or a good R-n-atom) if its measure is between $(\frac{1}{2})^{[E(R,S) \pm \beta/100]n}$, and if its n-name has the property that the distribution of names of length u in it is within $\bar{\delta}$ of dist $\bigvee_0^u S^{-i} R$ (dist $\bigvee_0^u S^{-i} R$ = dist of R-u-names under S).

Because of the Shannon-McMillan-Breiman theorem and the ergodic theorem, we can choose n so large that the measure of the union of atoms in $\bigvee_0^n T^{-i} Q$ and $\bigvee_0^n S^{-i} R$ that are not good is less than $\frac{\gamma}{100}$.

We can also choose n so large that $\frac{u}{n} < \frac{\gamma}{100}$, $n\beta > 100$, and $m(A) < \frac{1}{n}$ implies that $-[m(A) \log m(A) + (1 - m(A)) \log (1 - m(A))] < \frac{\gamma}{100}$.

There are more good R-n-atoms than there are good Q-n-atoms. Therefore, we can assign to each good Q-n-atom a good R-n-atom.

We can now apply the R-K theorem to obtain a set F such that $T^i F$, $0 \leq i \leq n$ are disjoint and $m\left(\bigcup_0^n T^i F\right) > 1 - \frac{\gamma}{100}$ and dist $\left(\bigvee_0^n T^{-i} Q/F\right)$ = dist $\left(\bigvee_0^n T^{-i} Q\right)$. Call the atoms in $\bigvee_0^n T^{-i} Q/F$ that are contained in good Q-n-atoms, good Q-n-F-atoms, and carry over our assignment of good R-n-atoms to good Q-n-F-atoms.

We will now define P' on $\bigcup_0^n T^i \bar{F}$ where $\bar{F}$ is the union of the good Q-n-F-atoms. If g is a good Q-n-F-atom, then $T^j g$ will lie in $(P')^i$ (the ith of P' atom) where i is the jth term in the R-n-name of the R-n-atom assigned to g. (Thus if g is a good Q-n-F-atom, then the P'-n-name of $x \in g$ will be the same as the R-n-name of the R-n-atom assigned to g.) It does not matter how we define P' on the rest of X. Let $\tilde{F}$ be the partition consisting of F and its complement. We have that $\bigvee_{-n}^{n} T^i(P' \vee \tilde{F})$, restricted to $\bigcup_0^n T^i\overline{F}$, refines Q. (We see this as follows: By the definition of Q-n-F-atoms we have that if g is a Q-n-F-atom and $0 \leq j \leq n$, then $T^j g$ is entirely contained in one atom of Q. Because each Q-n-F-atom in $\overline{F}$ was assigned a different R-n-atom we have that $\bigvee_0^n T^{-i}P'$, restricted to $\overline{F}$, is exactly the partition of $\overline{F}$ into Q-n-F-atoms. Therefore, $\bigvee_{-n}^{n} T^i(' \vee \tilde{F})$ contains all the $T^i g$, $0 \leq i \leq n$, g being a Q-n-F-atom in $\overline{F}$.) Because the measure of the union of good Q-n-atoms is $> 1 - \frac{\gamma}{100}$, and $m\left(\bigcup_0^n T^i F\right) > 1 - \frac{\gamma}{100}$, and because of our choice of γ, and because $E(\tilde{F}) < \frac{\gamma}{100}$, we get that $E(P',T) > E(R,S) - \overline{\delta}$.

Because the names of the good R-n-atoms have the property that the u-names in them are distributed well we get $\left|\text{dist} \bigvee_0^u T^i P' - \text{dist} \bigvee_0^u S^i R\right| < \delta$.

The following lemma and its corollary are used to get the full generality out of Lemma 7 (the main lemma of this section). They are used in reducing Lemma 7 to the case where $E(R,S) > E(P,T)$. If $S^i R$ are independent this can be done simply to changing P to $\overline{P}$ so that $|P - \overline{P}| < (\frac{\varepsilon}{100})^2$ and dist $\overline{P}$ = dist R. In any particular case when we want to apply Lemma 7 we have enough extra conditions to effect a simpler reduction, but since we apply Lemma 7 so often we pause here to get it in its full generality.

LEMMA 5. Assume R,S and P,T are such that $\overline{d}((R,S),(P,T)) < \gamma$. Then given u and δ we can find $\overline{P}$ such that: $|\overline{P}| = |P|$, $|\overline{P} - P| < 2\gamma$ and

$$\left|\text{dist}\left(\bigvee_0^u T^i \overline{P}\right) - \text{dist}\left(\bigvee_0^u S^i R\right)\right| < \delta.$$

<u>Proof.</u> It will be convenient to assume that $\delta < \gamma$:

(1) Pick n so that $\frac{u}{n} < \frac{1}{100}\delta$.

(2) Apply the R-K theorem to find F so that

(a) $T^i F$, $0 \leq i < n$ are disjoint,

(b) $m\left(\bigcup_{i=0}^{n-1} T^i F\right) > 1 - \frac{1}{100}\delta$,

(c) $\operatorname{dist}\left(\bigvee_0^n T^{-i}P/_F\right) = \operatorname{dist}\left(\bigvee_0^n T^{-i}P\right)$.

(3) Because $\bar{d}((R,S),(P,T)) < \gamma$ we can find partitions R_i of F such that

(a) $\operatorname{dist}\left(\bigvee_0^{n-1} R_i/_F\right) = \operatorname{dist}\left(\bigvee_0^{n-1} S^{-1}R\right)$,

(b) $\frac{1}{n}\Sigma|R_i/F - T^{-i}P/F| < \gamma$.

(4) Define $\overline{P}$ so that on T^iF, $\overline{P}$ and T^iR_i agree (it does not matter how we define $\overline{P}$ on the rest of X).

(5) Because of 3b and 2b we get $|\overline{P} - P| < 2\gamma$.

(6) It is easy to see (by 4 and 3a) that if $0 \leq i \leq n - u$, then

$$\operatorname{dist}\left(\bigvee_0^u T^{-j}\overline{P}/_{T^iF}\right) = \operatorname{dist}\left(\bigvee_0^u S^{-j}R\right).$$

(1) and (2) and (6) imply the conclusions of our lemma.

COROLLARY 6. Let R,S and P,T satisfy: $|R| = |P|$, R,S is F.D. and $\bar{d}((R,S),(P,T)) < \gamma$. Then given δ we can find $\overline{P}$ such that: $|\overline{P}| = |P|$, $|P - \overline{P}| < 2\gamma$, and either (a) $E(\overline{P},T) < E(R,S)$ or (b) $\bar{d}((\overline{P},T),(R,S)) < \delta$.

<u>Proof.</u> Because R,S is F.D., there exist u_1 and δ_1 so that if $\overline{P}$ satisfies (1) $|\overline{P}| = |P|$, (2) $\left|\operatorname{dist}\left(\bigvee_0^{u_1} T^i\overline{P}\right) - \operatorname{dist}\left(\bigvee_0^{u_1} S^iR\right)\right| < \delta_1$, and (3) $|E(R,S) - E(\overline{P},T)| < \delta_1$, then (b) $\bar{d}((\overline{P},T),(R,S)) < \delta$. If k is an integer, and if we let $u = ku_1$, and if $\hat{\delta}$ is > 0, and $< \delta_1$, then if $\overline{P}$ satisfied (2′) $\left|\operatorname{dist}\left(\bigvee_0^u T^i\overline{P}\right) - \operatorname{dist}\left(\bigvee_0^u S^iR\right)\right| < \hat{\delta}$, then $\overline{P}$ would also satisfy (2). Choose u and $\hat{\delta}$ so that (2′) implies (4) $E(\overline{P},T) - E(R,S) < \delta_1$. Now apply the previous lemma to get $\overline{P}$ such that $|P| = |\overline{P}|$, $|P - \overline{P}| < 2\gamma$, and $\overline{P}$ satisfies (2′). It is now clear that if $\overline{P}$ does not satisfy (a), then it will satisfy (b).

LEMMA 7. Let R,S and P,T satisfy the following: $|R| = |P|$. R,S is F.D., $E(R,S) \leq E(T)$ and (1) $\bar{d}[(R,S),(P,T)] < (\frac{\varepsilon}{100})^2$. Then given $\delta > 0$ we can find P' such that (1′) $\bar{d}[(R,S),(P',T)] < \delta$ and $|P - P'| < \varepsilon$.

The point of this lemma is the following: We are trying to find a model of R,S in T. We have already found a model P that is $\frac{\varepsilon^2}{100}$ good (in the $\bar{d}$-metric). We can then change P to P′ (by less than ε) to make it arbitrarily close to R,S in the $\bar{d}$-metric.

Proof of Lemma 7. By the previous corollary we can assume that $E(P,T) < E(R,S)$ and $\bar{d}[(R,S),(P,T)] < \frac{\varepsilon^2}{100}$. Because S,R is F.D., it will be enough to show that given $\bar{\delta}$ and u we can find P′ satisfying (1) $\left|\text{dist} \bigvee_0^u T^i P' - \text{dist} \bigvee_0^u S^i R\right| < \bar{\delta}$, and (2) $|E(P',T) - E(R,S)| < \bar{\delta}$ (and $|P - P'| < \varepsilon$).

Choose a refinement Q of P such that $E(R,S) - E(Q,T) = \beta > 0$ where $\beta < \frac{\bar{\delta}}{100}$. (This is where we use $E(P,T) < E(R,S)$.)

Choose $\gamma \leq \min(\bar{\delta},\varepsilon)$, and such that $|Q - Q| < \gamma$ implies that $E(Q',T) \geq E(Q,T) - \frac{\bar{\delta}}{100}$.

We will call an atom in $\bigvee_0^n T^{-i}Q$ good (or good Q-n-atoms) if its measure is between $(\frac{1}{2})^{[E(Q,T) \pm \beta/100]n}$.

We will call an atom in $\bigvee_0^n S^{-i}R$ good (or a good R-n-atom) if its measure is between $(\frac{1}{2})^{[E(R,S) \pm \beta/100]n}$, and if its n-name has the property that the distribution of u-names in it is within $\bar{\delta}$ of dist $\bigvee_0^u S^{-i}R$.

Because of the Shannon-McMillan-Breiman theorem and the ergodic theorem we can choose n so large that the measure of the union of atoms in $\bigvee_0^n T^{-i}Q$ and $\bigvee_0^n S^{-i}R$ that are not good is less than $\frac{\gamma}{100}$. We can also choose n so large that $\frac{u}{n} < \frac{\bar{\delta}}{100}$, $n\beta > 100$, and $m(A) < \frac{1}{n}$ implies that $-[m(A) \log m(A) + (1 - m(A)) \log(1 - m(A))] < \frac{\bar{\delta}}{100}$.

Because of hypothesis (1) we can choose partitions $\bar{R}_i$, and $0 \leq i \leq n$, such that dist $\bigvee_0^n \bar{R}_i$ = dist $\bigvee_0^n S^{-i}R$, and $\frac{1}{n+1} \sum_{i=0}^n |\bar{R}_i - T^{-i}P| < \frac{\varepsilon^2}{100}$. The good atoms in $\bigvee_0^n \overline{R}_i$ will be those that correspond to the good atoms in $\bigvee_0^n S^{-i}R$. We will call these good $\overline{R}$-n-atoms. The $\overline{R}$-n-name of x will be defined with respect to $\bigvee_0^n \overline{R}_i$ (the ith term being the atom of $\overline{R}_i$ contains x).

We will now pick out a subset of the good Q-n-atoms which we will call very good Q-n-atoms. First, let E be the set of x whose $\overline{R}$-n-name is good and differs from its P-n-name in $< \varepsilon n$ places. We will call g a very good Q-n-atom, if g is good and more than half of g lies in E. Because $\frac{1}{n+1} \sum_{i=0}^{n} |\overline{R}_i - T^{-i}P| < \frac{\varepsilon^2}{100}$, (note that the ith term in the $\overline{R}_i$ name of x is the index of the atom of R_i containing x, while the ith term in P-name of x is the index of the atom of $T^{-i}P$ containing x) we get that $m(E) > 1 - \varepsilon$ and hence the measure of the union of very good Q-n-atoms is $> 1 - \frac{\varepsilon}{2}$.

Because the measure of a good Q-n-atom is greater than two times the measure of a good $\overline{R}$-n-atom, we have that any ℓ very good Q-n-atoms intersect at least ℓ good $\overline{R}$-n-atoms in E. We can therefore apply the marriage lemma to assign to each very good Q-n-atom a good $\overline{R}$-n-atom which intersects it in E. Thus we have assigned to each very good Q-n-atom, g, a good $\overline{R}$-n-atom whose $\overline{R}$-n-name agrees with the P-n-name of any point in g, in more than $(1 - \frac{\varepsilon}{n})n$ places.

Because there are more good $\overline{R}$-n-atoms than there are good Q-n-atoms, we can extend the above assignment so that each good Q-n-atom is assigned a good $\overline{R}$-n-atom.

We can now apply the R-K theorem to obtain a set F such that T^iF, $0 \leq i \leq n$ are disjoint and $m\left(\bigcup_0^n T^iF\right) > 1 - \frac{\gamma}{100}$ and dist $\bigvee_0^n T^{-i}Q/F =$ dist $\bigvee_0^n T^{-i}Q$. Call the atoms in $\bigvee_0^n T^{-i}Q/F$ that are contained in good and very good Q-n-atoms, good and very good Q-n-F-atoms, and carry over our assignment of good $\overline{R}$-n-atoms to $\overline{Q}$-n-F-atoms.

We will now define P' on $\bigcup_0^n T^i\overline{F}$ where $\overline{F}$ is the union of the good Q-n-F-atoms. If g is a good Q-n-F-atom, then T^jg will lie in $(P')^i$ where i is the

j^{th} term in the $\overline{R}$-n-name of the $\overline{R}$-n-atom assigned to g. $((P')^i$ being the ith atom of P'.) It does not matter how we define P' on the rest of X.

Let $\tilde{F}$ be the partition consisting of F and its complement. Because each Q-n-F-atom in $\overline{F}$ was assigned a different R-n-aton, we have that $\bigvee_{-n}^{n} T^i(P' \vee \tilde{F})$ restricted to $\bigcup_{0}^{n} T^i\overline{F}$ refines Q. Because the measure of the union of good Q-n-atoms is $> 1 - \frac{\gamma}{100}$, $m(\bigcup_0^n T^i F) > \frac{\gamma}{100}$, because of our choice of γ, and because $E(\tilde{F}) < \overline{\delta}/100$, we get that $E(P',T) > E(R,S) - \overline{\delta}$.

Because the names of the good $\overline{R}$-n-atoms have the property that the u-names in them are distributed well, we get $\left| \text{dist} \bigvee_0^u T^i P' - \text{dist} \bigvee_0^u S^i R \right| < \overline{\delta}$. Lastly, since each very good Q-n-atom, g, has the property that the P-n-name of any point in g and the $\overline{R}$-n-name of the $\overline{P}$-n-atom assigned to g differ in less than $\frac{\varepsilon}{2}$n-places, and since the measure of the union of the very good Q-n-atoms is $> 1 - \frac{\varepsilon}{2}$, we have $|P' - P| < \varepsilon$. (Note that the $\overline{P}$-n-name of x in $g \cap F$ is the $\overline{R}$-n-name of the $\overline{R}$-n-atom assigned to g.)

PROPOSITION 8. Let R,S and P,T satisfy the following: $|R| = |P|$, R,S is F.D. $E(R,S) \leq E(T)$ and $\overline{d}[(R,S),(P,T)] < (\frac{\varepsilon}{100})^2$. Then we can find a P' such that $|P' - P| < \varepsilon$ and $P',T \sim R,S$.

<u>Proof</u>. This follows immediately, by applying Lemma 7 a countable number of times.

<u>Remark</u>. P' will in general not generate (if T is not Bernoulli, it could not possibly generate). The sequence of partitions that converge to P' can be chosen to come closer and closer to picking up a generator, but since it will take longer and longer to do so we do not get a generator in the limit.

PROPOSITION 9. Let R,S be F.D. Then given ε there is a δ and u such that if P,T satisfies (1) $E(T) \geq E(R,S)$, (2) $|P| = |R|$, (3) $\left| \text{dist}\left(\bigvee_0^u T^i P\right) - \text{dist}\left(\bigvee_0^u S^i R\right) \right| < \delta$, and (4) $|E(P,T) - E(R,S)| < \delta$, then there exists a P' such that (5) $|P - P'| < \varepsilon$, and (6) $P',T \sim R,S$.

<u>Proof</u>. Because of the definition of F.D., we can find δ and u so that (2), (3), and (4) imply $\overline{d}[(R,S),(P,T)] < \left(\frac{\varepsilon}{100}\right)^2$. Now apply Proposition 8.

Part III.

We will first introduce some new definitions.

DEFINITION. A "gadget" will be a collection of disjoint sets F_i, $0 \leq i \leq 1$, a map T which map F_i onto F_{i+1} in a measure-preserving invertible manner, (T is undefined on F_n) and a partition P of $\bigcup_0^n F_i$.

We say that the gadgets $G = \bigcup_0^n F_i$, P,T is isomorphic to $\overline{G} = \bigcup_0^n \overline{F}_i$, $\overline{P}$,$\overline{T}$, if $\operatorname{dist}\left[\left(\bigvee_0^n T^{-i}P\right)\Big/ F_0\right] = \operatorname{dist}\left[\left(\bigvee_0^n \overline{T}^{-i}\overline{P}\right)\Big/ \overline{F}_0\right]$. (That is, the distribution of the P-n-name in F_0, under T, is the same as the distribution of the $\overline{P}$-n-name in $\overline{F}_0$ under $\overline{T}$. If T is a transformation and F a set such that T^iF are disjoint for $0 \leq i \leq n$, and P is a partition of $\bigcup_0^n T^iF$, then $\bigcup_0^n T^iF$,P,T is a gadget. $\bigcup_0^n T^iF$,P,T is isomorphic to $\bigcup_0^n \overline{T}^i\overline{F}$,$\overline{P}$,$\overline{T}$ if and only if the distribution of the P-n-name in F, under T, is the same as the distribution of the $\overline{P}$-n-name in $\overline{F}$, under $\overline{T}$.

DEFINITION. Let L be a partition of the set of sequences $\{\alpha_i\}_{i=-K}^{K}$, $1 \leq \alpha_i \leq k$. Let P be a partition with k atoms. There is a 1 - 1 correspondence between the atom in $\bigvee_{-K}^{K} T^iP$ and sequences $\{\alpha_i\}_{i=-K}^{K}$. (Each atom in $\bigvee_{-K}^{K} T^iP$ has the form $\bigcap_{-K}^{K} T^iP^{\alpha_i}$.) $L\left(\bigvee_{-K}^{K} T^iP\right)$ is the partition (of which $\bigvee_{-K}^{K} T^iP$ is a refinement) whose i^{th} atom consists of the atoms in $\bigvee_{-K}^{K} T^iP$ whose corresponding sequences are in the ith atom of L.

LEMMA 10. Let T be a transformation and P and Q partitions such that P generates E(P,T) = E(Q,T) and P,T and Q,T are both F.D. Then given ε we can find $\overline{Q}$ such that (1) $\overline{Q}$,T $\sim$ Q,T, (2) $|Q - \overline{Q}| < \varepsilon$, and (3) $\bigvee_{-\infty}^{\infty} T^i\overline{Q} \supset^{\varepsilon} P$.

Proof. Our proof will divide into two parts. The first part will be devoted to showing:

(A) There is a $\overline{P}$ such that (i) $\overline{P} \subset \bigvee_{-\infty}^{\infty} T^i Q$, (ii) $\overline{P},T \sim P,T$, and (iii) there is a K and operator L such that $\left|L\left(\bigvee_{-K}^{K} T^i P\right) - Q\right| < \frac{1}{2}\varepsilon$ and $\left|L\left(\bigvee_{-K}^{K} T^i \overline{P}\right) - Q\right| < \frac{1}{2}\varepsilon$.

We get (A) as follows:

(1) Because P generates, there is a K and L such that (a) $\left|L\left(\bigvee_{-K}^{K} T^i P\right) - Q\right| < \frac{1}{100}\varepsilon$.

Applying Proposition 9 of Part II to T acting on $X_Q = \left(X, \bigvee_{-\infty}^{\infty} T^i Q, m\right)$, we get u and δ such that if P' satisfies (b) $|P'| = |P|$, (c) $\left|\operatorname{dist}\left(\bigvee_0^u T^i P'\right) - \operatorname{dist}\left(\bigvee_0^u T^i P\right)\right| < \delta$, and (d) $|E(P',T) - E(P,T)| < \delta$, then there is a $\overline{P}$ such that (e) $\overline{P} \subset \bigvee_{-\infty}^{\infty} T^i Q$, (f) $|\overline{P} - P'| < \frac{\varepsilon}{300K}$, and (g) $\overline{P},T \sim P,T$.

Choose $\overline{\varepsilon}$ such that $\overline{\varepsilon} < \delta$, $\overline{\varepsilon} < \varepsilon$, and if $|Q^*| = |Q|$ and $|Q^* - Q| < \overline{\varepsilon}$, then $|E(Q^*,T) - E(Q,T)| < \frac{1}{100}\delta$. Thus if P' is any partition such that $\bigvee_{-\infty}^{\infty} T^i P' \overset{\overline{\varepsilon}}{\supset} Q$, then $|E(P',T) - E(Q,T)| < \delta$.

We now choose K_1 such that ($K_1 > K$ and $K_1 > u$). (h) $\bigvee_{-K_1}^{K_1} T^i P \overset{\frac{1}{2}\overline{\varepsilon}}{\supset} Q$.

Choose N so that $K_1 < \frac{\overline{\varepsilon}}{100} N$.

(2) Apply the R-K theorem to find a set F such that $T^i F$, $0 \leq i \leq N$ are disjoint, and $m\left(\bigcup_0^N T^i F\right) > 1 - \frac{\overline{\varepsilon}}{100}$, and $F \subset \bigvee_{-\infty}^{\infty} T^i Q$.

Let G_1 be the gadget $\bigcap_0^N T^i F, P \vee Q, T$. Choose P' such that $P' \subset \bigvee_{-\infty}^{\infty} T^i Q$

and $G_2 = \bigcup_{0}^{N} T^iP, P' \vee Q, T$ is isomorphic to G_1 (extend the definition of P' to the rest of X in some arbitrary way).

(3) Because $G_1 \sim G_2$, we get (from 1(a) and 1(h)):

(a) $\left| L\left(\bigvee_{-K}^{K} T^iP'\right) - Q\right| < \frac{2}{100}\,\varepsilon$, and (b) $\bigvee_{-K_1}^{K_1} T^iP' \overset{\bar{\varepsilon}}{\supset} Q$.

(We get (a) and (b) as follows:

$\left| L\left(\bigvee_{-K}^{K} T^iP\right) - Q\right|$ is the measure of the set of x such that x is in i^{th} atom of Q, and its −K to K name $\{\alpha_i\}_{-K}^{K}$ lies in the jth atom of L and $i \neq j$. (Here $\alpha_i = k$ if T^ix lies in the kth atom of P.) Because $G_1 \sim G_2$, the contribution to $\left| L\left(\bigvee_{-K}^{K} T^iP'\right) - Q\right|$ coming from $\bigcup_{K}^{N-K} T^iF$ is less than $\frac{1}{100}\,\varepsilon$. Because $\bigcup_{K}^{N-K} T^iF$ is most of the space we get (a). The same argument gives (b).)

Because $G_1 \sim G_2$, we have that $\left|\text{dist}\left(\bigvee_{0}^{u} T^iP'\right) - \text{dist}\left(\bigvee_{0}^{u} T^iP\right)\right| \leq \bar{\varepsilon} \leq \delta$.

Because of 3(b) and the choice of $\bar{\varepsilon}$, we have $|E(P',T) - E(Q,T)| < \delta$.

Because of our choice of δ and u, we get $\bar{P}$ such that (c) $\bar{P} \subset \bigvee_{-\infty}^{\infty} T^iQ$,

(d) $\bar{P},T \sim P,T$, and (e) $|\bar{P} - P'| < \frac{\varepsilon}{100K}$. Because of (e) and (a) we have

(f) $\left| L\left(\bigvee_{-K}^{K} T^i\bar{P}\right) - Q\right| < \varepsilon$. (c), (d), and (f) give (A). We will now use (A) to prove our lemma (Q and $\bar{P}$ will act as a model for finding $\bar{Q}$). *

(4) Because $\bar{P} \subset \bigvee_{-\infty}^{\infty} T^iQ$, we can find M such that (a) $\bigvee_{-M}^{M} T^iQ \overset{\varepsilon/10}{\supset} \bar{P}$.

*In this part we will need quantities analogous to u, δ, $\bar{\varepsilon}$, N, F, G_1, and G_2. Although these will be different numbers we will denote them by the same letters. ε, K, and L will retain their old meaning.

As before, apply Proposition 9 of Part II to get a δ and u, such that if Q' satisfies (b) $|Q'| = |Q|$, (c) $\left|\text{dist} \bigvee_0^u T^i Q' - \text{dist} \bigvee_0^u T^i Q\right| < \delta$, and (d) $|E(Q',T) - E(Q,T)| < \delta$, then there is a $\overline{Q}$ satisfying (e) $|\overline{Q} - Q'| < \frac{\varepsilon}{300M}$, and (f) $\overline{Q},T \sim Q,T$.

Choose $\bar{\varepsilon}$ so that if $\bigvee_{-\infty}^{\infty} T^i Q' \overset{\bar{\varepsilon}}{\supset} P$, then $|E(Q',T) - E(P,T)| < \delta$.

Now choose M_1 so that $(M_1 > M, M_1 > u)$. (g) $\bigvee_{-M_1}^{M_1} T^i Q \overset{\frac{1}{2}\bar{\varepsilon}}{\supset} \overline{P}$.

Choose N so that $M_1 < \frac{\varepsilon}{100} N$.

(5) We now apply the R-K theorem to find F such that $T^i F$ are disjoint $0 \leq i \leq N$, $m\left(\bigcup_0^N T^i F\right) > 1 - \bar{\varepsilon}/100$, and (a) $\text{dist}\left(\bigvee_0^N T^i(\overline{P} \vee Q \vee P)_F\right) = \text{dist} \bigvee_0^N T^i(\overline{P} \vee Q \vee P)$. Let G_1 be the gadget, $\bigcup_0^N T^i F, \overline{P} \vee Q$. Define Q' so that the gadget $G_2 = \bigcup_0^N T^i F, P \vee Q'$ is isomorphic to G_1. (This can be done because the gadgets $\bigcup_0^N T^i F, P$ and $\bigcup_0^N T^i F, \overline{P}$ are isomorphic by (a), and the fact that $\overline{P},T \sim P,T$.)

(6) Since $G_1 \sim G_2$, 4(a) and 4(g) imply (a) $\bigvee_{-M}^{M} T^i Q' \overset{\varepsilon/5}{\supset} P$, and (b) $\bigvee_{-M_1}^{M_1} T^i Q' \overset{\bar{\varepsilon}}{\supset} P$

Because $G_1 \sim G_2$, $\left|\text{dist}\left(\bigvee_0^u T^i Q'\right) - \text{dist} \bigvee_0^u T^i Q\right| < \delta$, and because of (b) and our choice of $\bar{\varepsilon}$, $|E(Q',T| - E(P,T)| < \delta$. Because of our choice of δ and u we can find a $\overline{Q}$ satisfying (c) $|\overline{Q} - Q'| < \frac{\varepsilon}{30M}$, and (d) $\overline{Q},T \sim Q,T$.

Because of (c) and (a) we get (e) $\bigvee_{-M}^{M} T^i \overline{Q} \overset{\varepsilon}{\supset} P$.

Because $G_1 \sim G_2$, and because (f) $\left|L\left(\bigvee_{-K}^{K} T^i \overline{P}\right) - Q\right| < \frac{1}{100}\varepsilon$ (see (A)), we get that $\left|L\left(\bigvee_{-K}^{K} T^i P\right) - Q'\right| < \frac{2}{100}\varepsilon$, and because $|Q' - \overline{Q}| < \frac{\varepsilon}{30M}$, we get

(g) $\left| L\left(\bigvee_{-K}^{K} T^iP\right) - \overline{Q}\right| < \frac{1}{2}\varepsilon$. However, (h) $\left| L\left(\bigvee_{-K}^{K} T^iP\right) - Q\right| < \frac{1}{2}\varepsilon$ (see (A)).

Because of (g) and (h) we have (i) $|\overline{Q} - Q| < \varepsilon$.

Our lemma follows from (d), (e), and (i).

PROPOSITION 11. Let T be a transformation and P and Q partitions such that P generates, E(P,T) = E(Q,T) and (P,T) and (Q,T) are F.D. Then given ε there is a $\overline{Q}$ such that $|Q - \overline{Q}| < \varepsilon$, $\overline{Q}$ generates and $\overline{Q},T \sim Q,T$.

Proof. This follows easily from the previous lemma. Form Q_1 by the previous lemma so that (a) $Q_1,T \sim Q,T$, (b) $|Q_1 - Q| < \varepsilon_1$, and (c) $\bigvee_{-N_1}^{N_1} T^iQ_1 \overset{\varepsilon_1}{\supset} P$. We apply the lemma again to get Q_2, where $|Q_2 - Q_1| < \varepsilon_2$, and ε_2 is so small that $\bigvee_{-N_1}^{N_1} T^iQ_2 \overset{(1+\frac{1}{2})\varepsilon_1}{\supset} P$, and $\bigvee_{-N_2}^{N_2} T^iQ_2 \overset{\varepsilon_2}{\supset} P$. Applying the lemma once more, we get Q_3 such that $\bigvee_{-N_1}^{N_1} T^iQ_3 \overset{(1+\frac{1}{2}+\frac{1}{2^2})\varepsilon_1}{\supset} P$, $\bigvee_{-N_2}^{N_2} T^iQ_3 \overset{(1+\frac{1}{2})\varepsilon_2}{\supset} P$, and $\bigvee_{-N_3}^{N_3} T^iQ_3 \overset{\varepsilon_3}{\supset} P$. Continuing in this way we get that if the ε_i are chosen so that in addition, $\varepsilon_i < \frac{1}{2^i}$, then the Q_i converge to a $\overline{Q}$ and $\overline{Q},T \sim Q \sim T$. Furthermore, $\bigvee_{-N_1}^{N_1} T^i\overline{Q} \overset{2\varepsilon_1}{\supset} P$, $\bigvee_{-N_2}^{N_2} T^i\overline{Q} \overset{2\varepsilon_2}{\supset} P$, etc. Therefore, $\overline{Q}$ will generate.

THEOREM 12. If E(T) = E(S), and if T and S have finite partitions which generate and are F.D., then T is isomorphic to S.

Proof. Let P be the generator of T, then by Part II and III there is a generator $\overline{P}$ of S, and $\overline{P},S \sim P,T$. Thus S and T are isomorphic in the measure algebra sense (see Appendix 1). (If the spaces they act on are Lebesgue spaces, then they are also isomorphic in the sense of the Introduction.)

THEOREM 13. Two Bernoulli shifts with the same entropy are isomorphic.

Proof. This is a special case of the previous theorem, because of Part I. By the definition of Bernoulli shift the spaces involved are Lebesgue spaces.

The isomorphism theorem gives us information about a Bernoulli shift because it allows us to represent the shift in different ways. For example, we have:

COROLLARY. Bernoulli shifts have roots of all orders.

Proof. Let T be a Bernoulli shift. Let T_1 be a Bernoulli shift such that $E(T_1) = \frac{1}{k}E(T)$. It is easy to see that $E(T_1^k) = k\,E(T_1)$. Hence $T_1^k \sim T$. Hence T has a kth root.

Remark. It was not previously known if the 2-shift had a square root.

Remark. We will show later that Bernoulli shifts can even be imbedded in flows.

5. Factors of Bernoulli shifts

In this section we will characterize the P,T such that T is Bernoulli. We will show that if T is Bernoulli, then for any finite partition P, P,T is finitely determined, and we have already seen that if P,T is finitely determined, then T acting on $\bigvee_{-\infty}^{\infty} T^i P$ is a Bernoulli shift. This will imply that factors of Bernoulli shifts are Bernoulli shifts. We will also show that roots of Bernoulli shifts are necessarily Bernoulli.

The following lemma is put in for expository purposes. It is a simple case of the main theorem of this section and should make the proof of that theorem easier to follow.

LEMMA 1. Let T be a Bernoulli shift and P a finite partition such that $E(P,T) = E(T)$. Then given ε, there is a δ and n such that if $\overline{P},\overline{T}$ satisfies, $\overline{T}$ is ergodic, $E(\overline{T}) \geq E(T)$, $|P| = |\overline{P}|$, and (1) $|E(\overline{P},\overline{T}) - E(P,T)| < \delta$, and (2) $\left|\text{dist}\left(\bigvee_0^n T^i P\right) - \text{dist}\left(\bigvee_0^n \overline{T}^i \overline{P}\right)\right| < \delta$. Then there is a $\hat{P}$ such that: (1) $|\hat{P} - \overline{P}| < \varepsilon$, and (2) $\hat{P},\overline{T} \sim P,T$.

Before giving the proof, we should recall the definition of $L\left(\bigvee_{-K}^{K} T^i P\right)$ given at the beginning of section 4, Part II.

Proof. Let B be a finite partition such that $T^i B$ are independent and generate. Pick K so the $\bigvee_{-K}^{K} T^i B \overset{1/10\varepsilon}{\supset} P$. Let L be as in the above definition and such that $\left|L\left(\bigvee_{-K}^{K} T^i B\right) - P\right| < \frac{1}{10}\varepsilon$. Pick δ' (using Proposition 9 and Corollary 3, section 4) such that if $|B'| = |B|$ if and only if $|E(B',\overline{T}) - E(B,T)| < \delta'$, and $|\text{dist } B' - \text{dist } B| < \delta'$, then there is a $\overline{B}$ such that $|\overline{B} - B'| < \frac{\varepsilon}{100K}$, $\text{dist }(\overline{B}) = \text{dist }(B)$, and $\overline{T}^i\overline{B}$ are independent. Now pick γ, $\gamma < \frac{1}{100}\varepsilon$, and $\gamma < \frac{1}{100}\delta'$ such that if $|P_1| = |\overline{P}|$ and $|P_1 - \overline{P}| < \gamma$, then $|E(P_1,\overline{T}) - E(\overline{P},\overline{T})| < \frac{1}{10}\delta'$. Now pick $K' > K$ so that $\bigvee_{-K'}^{K_1} T^i B \overset{1/10\gamma}{\supset} P$. Pick n so that $\frac{K'}{n} < \frac{1}{100}\gamma$ and $\left|\frac{1}{n}\log\frac{1}{n} + (1 - \frac{1}{n})\log(1 - \frac{1}{n})\right| < \frac{1}{10}\delta'$. Note that so far our choice of n and γ depended only on T,P, and B. We now choose δ so that $\delta < \frac{\gamma}{100}$. Having chosen n and δ we will now assume that we have $\overline{P},\overline{T}$ satisfying hypothesis (0), (1), and (2).

Now apply the R-K theorem to find a set F such that $T^i F$, $0 \leq i \leq n - 1$, are disjoint, and $m\left[X - \bigcup_{i=0}^{n-1} T^i F\right] < \frac{1}{100}\gamma$ (X is the space on which T acts) and (1) $\text{dist}\left(\bigvee_0^{n-1} T^{-i}(P \vee B)/F\right) = \text{dist}\left(\bigvee_0^{n-1} T^{-i}(P \vee B)\right)$.

Applying the Rochlin's theorem again, we can find a set $\overline{F}$ such that $\overline{T}^i\overline{F}$,

$0 \le i \le n-1$, are disjoint $m\left(\overline{X} - \bigcup_{i=0}^{n-1} \overline{T}^i\overline{F}\right) < \frac{\gamma}{100}$, and

(2) $\text{dist}\left(\bigvee_0^{n-1} \overline{T}^i\overline{P}/\overline{F}\right) = \text{dist}\left(\bigvee_0^{n-1} \overline{T}^i\overline{P}\right)$. Now because of hypothesis (2) and our choice of n and δ, we could (by removing $< \delta\ m(\overline{F})$ from $\overline{F}$) assume that we have instead of (2) the following:

(3) $\text{dist}\left(\bigvee_0^{n-1} \overline{T}^i\overline{P}/\overline{F}\right) = \text{dist}\left(\bigvee_0^{n-1} T^{-i}P\right)$.

If we let G be the gadget formed by partitioning T^iF by P, and $\overline{G}$ the gadget formed by partitioning the $\overline{T}^i\overline{F}$ by $\overline{P}$, then G and $\overline{G}$ are isomorphic because of (1) and (3). Let G' be the gadget formed by partitioning T^iF by $P \vee B$. Choose a partition B' of $\bigcup_{i=0}^{n-1} \overline{T}^i\overline{F}$ such that if we form the gadget $\overline{G}'$ by partitioning the $\overline{T}^i\overline{F}$ by $\overline{P} \vee B'$, then $\overline{G}'$ is isomorphic to G'. (Now extend B' to the rest of the space in any way.)

Because $\overline{G}'$ and G' are isomorphic and fill up most of the space, because $\frac{K}{n} < \frac{1}{10}\gamma$, and because $\left|L\left(\bigvee_{-K}^{K} T^iB\right) - P\right| < \frac{1}{10}\varepsilon$, we get that

(4) $\left|L\left(\bigvee_{-K}^{K} \overline{T}^iB'\right) - \overline{P}\right| < \frac{2}{10}\varepsilon$.

We also have $\bigvee_{-K'}^{K'} T^iB' \overset{2/10\gamma}{\supset} \overline{P}$, and hence by our choice of γ,

(5) $|E(B',\overline{T}) - E(\overline{P},\overline{T})| < \frac{1}{10}\delta'$. Furthermore, since $\overline{G}'$ and G' are isomorphic and fill up most of the space, we get that (6) $|\text{dist}(B') - \text{dist}(B)| < \delta'$. (5), (6), and our choice of δ' (recalling that $|E(\overline{P},\overline{T}) - E(B,T)| < \delta < \delta'$) imply that there is a $\overline{B}$ such that: (7) $|\overline{B} - B'| < \frac{\varepsilon}{100K}$ and (8) $\overline{B},\overline{T} \sim B,T$. Because of (4) and (7), we get that (9) $\left|L\left(\bigvee_{-K}^{K} \overline{T}^i\overline{B}\right) - \overline{P}\right| < \frac{2}{10}\varepsilon$. Because of (8), $\overline{T}$ acting on $\bigvee_{-\infty}^{\infty} \overline{T}^i\overline{B}$ is isomorphic to T under an isomorphism φ, of the measure algebras which takes B on $\overline{B}$. Let $\hat{P}$ be $\varphi(P)$. Then $P,T \sim \hat{P},T$. Again, because of the isomorphism, and because L was defined so

that $\left|L\left(\bigvee_{-K}^{K} T^iB\right) - P\right| < \frac{1}{10}\varepsilon$, we get that (10) $\left|L\left(\bigvee_{-K}^{K} \overline{T}^i\overline{B}\right) - \hat{P}\right| < \frac{1}{10}\varepsilon$.

(9) and (10) imply $|\overline{P} - \hat{P}| < \varepsilon$. We will now modify the proof of the previous lemma to cover the case where $E(P,T) \leq E(T)$.

THEOREM 2. Let T be a Bernoulli shift (of finite entropy), and let P be a finite partition. Then, P,T is finitely determined.

Proof. We will show that P,T is finitely determined by proving the following: (A) Let $\hat{T}$ be a Bernoulli shift acting on $\hat{X}$ satisfying $E(\hat{T}) = E(T)$. Then given ε there is a δ and an n such that if $\overline{T}$ is ergodic and acts on $\overline{X}$, and if (0) $|\overline{P}| = |P|$, (1) $|E(\overline{P},\overline{T}) - E(P,T)| < \delta$. (2) $\left|\text{dist}\left(\bigvee_0^n \overline{T}^i\overline{P}\right) - \text{dist}\left(\bigvee_0^n T^iP\right)\right| < \delta$, then there is a partition $\tilde{P}$ of $X \times \overline{X}$ such that $(\tilde{P},\overline{T} \times \hat{T}) \sim P,T$ and $|\tilde{P} - (\overline{P} \times \hat{X})| < \varepsilon$. ($\overline{P} \times \hat{X}$ is the partition of $\overline{X} \times \hat{X}$ formed by partitioning the $\overline{X}$ coordinate by $\overline{P}$. Thus $\overline{P} \times \hat{X}, \overline{T} \times \hat{T} \sim \overline{P},\overline{T}$.)

We will now prove (A). Let B be an independent generator for T. Pick K so that $\bigvee_{-K}^{K} T^iB \overset{1/10\varepsilon}{\supset} P$. Define L so that $\left|L\left(\bigvee_{-K}^{K} T^iB\right) - P\right| < \frac{1}{10}\varepsilon$.

Pick δ_1 (using Proposition 9 and Corollary 3 of section 4) so that if T_1 is ergodic, $E(T_1) \geq E(T)$ and $|B_1| = |B|$, $|E(B_1,T_1) - E(B,T)| < \delta_1$ and $|\text{dist } B_1 - \text{dist } B| < \delta_1$, then there is $\overline{B}$ such that $\overline{B},T_1 \sim B,T$ and $|\overline{B} - B_1| < \frac{\varepsilon}{100K}$. Pick γ such that $\gamma < \frac{1}{100}\varepsilon$ and $\gamma < \frac{1}{100}\delta_1$, and if P_2 and P_3 are any two partitions of the same space such that $|P_2| \leq |\overline{P}| \cdot |B|$, $|P_3| \leq |\overline{P}| \cdot |B|$ and $|P_2 - P_3| < \gamma$, then there is an R such that $E(R) < \frac{1}{100}\delta_1$ and $P_2 \vee R \supset P_3$. Pick K' such that $\bigvee_{-K'}^{K'} T^iB \overset{1/10\gamma}{\supset} P$.

Let $\hat{T}$ be a Bernoulli shift such that $E(\hat{T}) > E(T)$. Let $\hat{Q}$ be an independent partition under $\hat{T}$ such that $|\hat{Q}| \leq |B|$, and $E(T) - \frac{1}{10}\delta_1 < E(\hat{Q}) + E(P,T) < E(T) - \frac{1}{100}\delta_1$.

If n is large enough, then the Shannon-McMillan-Brieman theorem implies that there is a collection C_1 of atoms of $\bigvee_{-n+1}^{0} \hat{T}^i\hat{Q}$, a collection C_2 of atoms of $\bigvee_{-n+1}^{0} T^i(P \vee B)$, and a collection $\overline{C}_2$ of atoms of $\bigvee_{-n+1}^{0} T^iP$ such that (a) $m(C_1) > 1 - \frac{1}{100}\gamma$, $m(C_2) > 1 - \frac{1}{100}\gamma$, $m(\overline{C}_2) > \frac{1}{100}\gamma$, and (b) any atom in C_2 is contained in an atom in $\overline{C}_2$.

(c) If $\overline{A}_2 \in \overline{C}_2$, then $m(\overline{A}_2) > \frac{1}{2}^{[E(P,T) + \frac{1}{1000}\delta_1]n}$,

if $A_1 \in C_1$, then $m(A_1) > \frac{1}{2}^{[E(\hat{T}) + \frac{1}{1000}\delta_1]n}$,

if $A_2 \in C_2$, then $m(A_2) < \frac{1}{2}^{[E(T) - \frac{1}{1000}\delta_1]n}$,

thus if n is large enough we will have: (B) $\frac{1}{100}\gamma\ m(A_1)\, m(\overline{A}_2) > m(A_2)$.

Pick an n so large, that in addition to the above conditions we have $\frac{K'}{n} < \frac{1}{100}\gamma$ and $|\frac{1}{n}\log\frac{1}{n} + (1-\frac{1}{n})\log(1-\frac{1}{n})| < \frac{1}{100}\delta_1$. Choose $\delta < \frac{\gamma}{100}$.

We have now chosen δ and n. Let us now assume we have a $\overline{P},\overline{T}$ satisfying the conditions in (A). Apply the R-K theorem to obtain an F such that T^iF, $0 \le i \le n-1$ are disjoint and fill up X to within $\frac{1}{100}\gamma$, and

$$\operatorname{dist}\left(\bigvee_{-n+1}^{0} T^i(P \vee B)\right) = \operatorname{dist}\left[\bigvee_{-n+1}^{0} T^i(P \vee B)\Big/F\right].)$$

Applying the R-K theorem again and using condition 2 of (A) we get $\overline{F}$ such that $\overline{T}^i\overline{F}$, $0 \le i \le n-1$ are disjoint, fill up $\overline{X}$ to within $\frac{1}{100}\gamma$ and

$$\operatorname{dist}\left(\bigvee_{-n+1}^{0} \overline{T}^i\overline{P}\Big/\overline{F}\right) = \operatorname{dist}\left(\bigvee_{-n+1}^{0} T^iP\right).$$

Let $P_1 = \overline{P} \times \hat{X}$, $F_1 = \overline{F} \times \hat{X}$, and let $T_1 = \overline{T} \times \hat{T}$, $X_1 = \overline{X} \times \hat{X}$, and $Q_1 = \hat{Q} \times \overline{X}$. (We thus have: $T_1^{\,i}F_1$ are disjoint, fill up X_1 to within $\frac{1}{100}\gamma$, and

$$\operatorname{dist}\left(\bigvee_{-n+1}^{0} T_1^{\,i}P_1\Big/F_1\right) = \operatorname{dist}\left(\bigvee_{-n+1}^{0} T^iP\right).)$$

Let G be the gadget $\bigcup_{i=0}^{n-1} T^i F$ partitioned by P and $\tilde{G}$, and $\bigcup_{i=0}^{n-1} T^i P$ partitioned by $P \vee B$. Let G_1 be the gadget $\bigvee_0^{n-1} T_1{}^i F_1$ partitioned by P_1.

Since $G \sim G_1$, we can find B_1 such that if $\tilde{G}_1$ is formed by $\bigcup_0^{n-1} T_1{}^i F_1$ partitioned by $P_1 \vee B_1$, then $\tilde{G} \sim \tilde{G}_1$. All that is involved in checking that $\tilde{G} \sim \tilde{G}_1$ is the measure of atoms of $\bigvee_{-n+1}^{0} T_1{}^i (B_1 \vee P_1) \big/ F_1$. We are still free to arrange the atoms of $\bigvee_{-n+1}^{0} T_1{}^i (B_1 \vee P_1) \big/ F_1$ within the atoms of $\bigvee_{-n+1}^{0} T_1{}^i (P_1) \big/ F_1$ in any way we want.

We will now show that B_1 could have been chosen with the additional property that $\bigvee_{-n}^{n} T_1{}^i (B_1 \vee P_1 \vee F_1) \overset{\frac{1}{10}\gamma}{\supset} Q_1$. (Here F_1 denotes the partition consisting of F_1 and its complement.) Because of (B) we have that if A_1 is in $\bigvee_{-n+1}^{0} T_1{}^i Q_1 \big/ F_1$ and corresponds to an atom in C_1, and A_2 is an atom in $\bigvee_{-n+1}^{0} T_1{}^i (P_1 \vee B_1) \big/ F_1$ and corresponds to an atom in C_2 and $\overline{A}_2$, the atom of $\bigvee_{-n+1}^{0} T_1{}^i P_1 \big/ F_1$ containing A_2, then $\frac{1}{100}\gamma \cdot m(A_1 \cap \overline{A}_2) > m(A_2)$. This means that we can change B_1 so that all but $\frac{\gamma}{50}$ of the atoms of $\bigvee_{-n+1}^{0} T_1{}^i (P_1 \vee B_1) \big/ F_1$ that correspond to atoms in C_2 lie inside an atom of $\bigvee_{-n+1}^{0} T_1{}^i Q_1 \big/ F_1$. Because $m(C_2) > 1 - \frac{1}{100}\gamma$ we get that

$$\bigvee_{-n}^{n} T_1{}^i (B_1 \vee P_1 \vee F_1) \overset{\frac{1}{10}\gamma}{\supset} Q_1 .$$

Since $\tilde{G} \sim \tilde{G}_1$ we could define B_1 off $\bigcup_0^{n-1} T_1^{\,i} F_1$ so that dist B = dist B_1. We will now show that $|E(B_1,T_1) - E(B,T)| < \delta_1$. (Since $E(B,T) = E(B) = E(B_1) \geq E(B_1,T_1)$, it is enough to show that $E(B_1,T_1) \geq E(B,T) - \delta_1$.)

Because $\tilde{G}_1 \sim \tilde{G}$, because $\bigvee_{-K'}^{K_1} T^i B \overset{\frac{1}{10}\gamma}{\supset} P$, and because $\tilde{G}$ and G_1 fill up most of X and X_1, we get that $\bigvee_{-K'}^{K'} T_1^{\,i} B_1 \overset{\gamma}{\supset} P_1$. Because of our choice of γ, there is a partition R, $E(R) < \frac{1}{100}\delta_1$, such that $\bigvee_{-K'}^{K'} T_1^{\,i}(B_1 \vee R) \supset P_1$. Because $\bigvee_{-n}^{n} T_1^{\,i}(B_1 \vee P_1 \vee F_1) \overset{\frac{1}{10}\gamma}{\supset} Q_1$, we have $\bigvee_{-\infty}^{\infty} T_1^{\,i}(B_1 \vee R \vee F_1) \supset P_1 \vee Q_1$. Thus $E(B_1 \vee R \vee F_1, T_1) \geq E(\hat{Q},\hat{T}) + E(\overline{P},\overline{T}) - \frac{1}{10}\delta_1 \geq E(T) - \frac{2}{10}\delta_1$. Hence $E(B_1,T_1) \geq E(T) - \delta_1$.

Because of our choice of δ_1 there is a $\overline{B}$ such that $|\overline{B} - B_1| < \frac{\varepsilon}{100K}$ and $\overline{B},T_1 \sim B,T$. Let φ be the isomorphism between T_1 acting on $\bigvee_{-\infty}^{\infty} T_1^{\,i}\overline{B}$, and T acting on $\bigvee_{-\infty}^{\infty} T^i B$ that takes B onto $\overline{B}$. Let $\tilde{P} = \varphi(P)$. (Thus $\tilde{P},T_1 \sim P,T$.)

Since $\tilde{G} \sim \tilde{G}_1$, and since $\left| L\left(\bigvee_{-K}^{K} T^i B\right) - P\right| < \frac{1}{10}\varepsilon$, we get $\left| L\left(\bigvee_{-K}^{K} T_1^{\,i} B_1\right) - P_1\right| < \frac{2}{10}\varepsilon$, and because $|B_1 - \overline{B}| < \frac{\varepsilon}{100K}$, we get $\left| L\left(\bigvee_{-K}^{K} T_1^{\,i} \overline{B}\right) - P_1\right| < \frac{3}{10}\varepsilon$. On the other hand, $\left| \bigvee_{-K}^{K} (T_1^{\,i}\overline{B}) - \tilde{P}\right| < \frac{1}{10}\varepsilon$ (since $\varphi(B) = \overline{B}$ and $\varphi(P) = \tilde{P}$). Thus $|P_1 - \tilde{P}| < \varepsilon$. This gives (A) (recall that $P_1 = \overline{P} \times \hat{X}$, and $T_1 = \overline{T} \times \hat{T}$).

COROLLARY 3. Factors of Bernoulli shifts (of finite entropy) are Bernoulli shifts.

Remark. A more exact statement of the above corollary is: Let T be a

Bernoulli shift such that $E(T) < \infty$. Let P be a finite partition. Then there is a finite partition B in $\bigvee_{-\infty}^{\infty} T^i P$ such that $T^i B$ are independent and $\bigvee_{-\infty}^{\infty} T^i B = \bigvee_{-\infty}^{\infty} T^i P$.

As a simple application of the F.D. characterization of arbitrary partitions under a Bernoulli shift we have:

THEOREM 4. If T^k is a Bernoulli shift then T is a Bernoulli shift.

Proof. Let P be a finite generator for T^k, and hence for T. We must show that given ε we can find δ and u such that if (a) $|\overline{P}| = |P|$, (b) $\left|\operatorname{dist}\left(\bigvee_1^{kn} \overline{T}^i \overline{P}\right) - \operatorname{dist}\left(\bigvee_1^{kn} T^i P\right)\right| < \delta$, and (c) $|E(P,T) - E(\overline{P},\overline{T})| < \delta$, then (d) $\overline{d}((P,T),(\overline{P},\overline{T})) < \varepsilon$, (b) implies (b′) $\left|\operatorname{dist}\left(\bigvee_1^{n} (\overline{T}^k)^i \left(\bigvee_1^{k} \overline{T}^j \overline{P}\right)\right) - \operatorname{dist}\left(\bigvee_1^{n} (\overline{T}^k)^i \left(\bigvee_1^{k} \overline{T}^j \overline{P}\right)\right)\right| < \delta$, and (c) implies (c′) $\left|E\left(\left(\bigvee_1^{k} \overline{T}^j \overline{P}\right), \overline{T}^k\right) - E\left(\left(\bigvee_1^{k} T^j P\right), T^k\right)\right| < k\delta$. Because T^k is Bernoulli $\bigvee_1^{k} T^i P, T^k$ is F.D. Thus δ and u can be chosen so that if $\overline{P},\overline{T}$ is such that (b′) and (c′) are satisfied (and $|\overline{P}| = |P|$), then (e) $\overline{d}\left(\left(\bigvee_1^{k} T^i P, T^k\right), \left(\bigvee_1^{k} \overline{T}^i \overline{P}, \overline{T}^k\right)\right) < \varepsilon$. But (e) implies $\overline{d}((P,T),(\overline{T},\overline{P})) < \varepsilon$.

6. An application to stationary processes

DEFINITION. Let T be a Bernoulli shift and P an arbitrary (finite) partition. We will say that the process P,T is a Bernoulli process or B-process.

THEOREM 1. Let P_i, T_i and P,T satisfy: T_i are Bernoulli shifts, and $\lim_{i \to \infty} \overline{d}((P_i,T_i),(P,T)) = 0$. Then T acting on $\bigvee_{-\infty}^{\infty} T^i P$ is a Bernoulli shift.

Remark. The above theorem says that the B-processes are closed in the $\bar{d}$-metric.

Proof. By Theorem 2, section 5, we get that P_i, T_i are finitely determined. (We can also assume that $|P_i| = |P|$.)

Let $\bar{T}$ be a Bernoulli shift such that $E(\bar{T}) \geq \sup E(P_i, T_i)$.

The proof will consist of showing that there is a $\bar{P}$ such that $\bar{P}, \bar{T} \sim P, T$. (Because of Corollary 3, section 5, $\bar{T}$ acting on $\bigvee_{-\infty}^{\infty} T^i \bar{P}$ is Bernoulli.)

By taking a subsequence of the P_i, T_i, we can assume that

$$\sum_{1}^{\infty} [\bar{d}((P_i, T_i), (P, T))]^{\frac{1}{2}} < \infty .$$

Because P_1, T_1 is F.D. and $E(\bar{T}) \geq E(P_1, T_1)$, there is (by Lemma 4 and Proposition 8 of section 4) a $\bar{P}_1$ such that $\bar{P}_1, \bar{T} \sim P_1, T_1$. Because P_2, T_2 is F.D., we can apply Proposition 8 of section 4 to get $\bar{P}_2$ such that

$$|\bar{P}_2 - \bar{P}_2| < 100\, [\bar{d}((P_1, T_1), (P_2, T_2))]^{\frac{1}{2}}, \text{ and } \bar{P}_2, \bar{T} \sim P_2, T_2 .$$

In general, if we have $\bar{P}_i$ such that $\bar{P}_i, \bar{T} \sim P_i, T_i$, then we can find $\bar{P}_{i+1}$ such that $|\bar{P}_i - \bar{P}_{i+1}| < 100\, [\bar{d}\left((P_i, T_i), (P_{i+1}, T_{i+1})\right)]^{\frac{1}{2}}$ and $\bar{P}_{i+1}, \bar{T} \sim P_{i+1}, T_{i+1}$.

It is now easy to see that the $\bar{P}_i$ will converge to a partition $\bar{P}$ and $\bar{P}, \bar{T} \sim P, T$ (i.e., $\bar{d}((\bar{P}, \bar{T}), (P, T)) = 0$).

DEFINITION. A finite coding (of length ℓ) of a process P,T will be another process obtained as follows: Divide the sequences $\{\alpha_i\}_{i=-\ell}^{\ell}$ (where the α_i are possible outputs of P,T) into k classes and let $f(\{\alpha_i\}_{-\ell}^{\ell})$ denote that class. This gives us a map from sequences $\{\alpha_i\}_{-\infty}^{\infty}$ to sequences $\{\beta_i\}_{-\infty}^{\infty}$ defined by $\beta_i = f(\{\alpha_j\}_{i-\ell}^{i+\ell})$. We also get a measure on on the sequences $\{\beta_i\}_{-\infty}^{\infty}$ by giving a set of $\{\beta_i\}$ sequences the measure of its inverse image. This gives us another process which we will call a finite coding of P,T.

Physically, a finite coding of P,T is obtained by looking at the last $2\ell + 1$

outputs of P,T and then printing out something depending in a fixed way on what one saw.

COROLLARY 2. The B-processes are the closure in the $\bar{d}$-metric of finite codings of independent processes.

Proof. This is obvious, since if P,T is a B-process, then there is an independent generator Q such that for each ε there is an ℓ and $Q \overset{\varepsilon}{\supset} \bigvee_{-\ell}^{\ell} T^i Q$.

Remark. This means that the B-processes are exactly those processes that can be approximated arbitrarily well in the $\bar{d}$-metric by a finite coding of a roulette wheel.

We have shown that if T is Bernoulli, then P,T is finitely determined. Thus:

THEOREM 3. If a sequence of (ergodic) processes P_i, T_i converges to a B-process P,T in the vague topology and $E(P_i, T_i) \to E(P,T)$, and P, P_i all have the same number of atoms, then the sequence converges in the $\bar{d}$-metric. Furthermore, the B-processes are the only processes for which the above holds.

Remark. The above theorem says that a B-process, P,T, is determined to within ε in the $\bar{d}$-metric by a finite number of parameters. That is, given ε there is a K and δ such that if Q,S is any other ergodic process whose entropy is within δ of the entropy of P,T, and such that the measure Q,S gives to sequences of length K is within ε of the measure that P,T gives to sequences of length K, then $\bar{d}((P,T),(Q,S)) < \varepsilon$.

This theorem can be viewed as a measure theoretic analog of structural stability.

7. A criterion for showing specific transformations are Bernoulli

In this section we will define a property of a process P,T called "very weak Bernoulli" or V.W.B. and show that if P,T is V.W.B., then T acting on $\bigvee_{-\infty}^{\infty} T^i P$ is a Bernoulli shift.*

DEFINITION. P,T is V.W.B. If given ε, then (A): there is an n such that for all m we can find a collection C_m of atoms of $\bigvee_{-m}^{-0} T^i P$ such that $m(C_m) > 1-\varepsilon$, and if $A \in C_m$, then $\bar{d}(\{T^i P\}_1^n, \{T^i P/A\}_1^n) < \varepsilon$.

LEMMA 1. Assume P,T is V.W.B. Then given ε, there is a u and δ such that if $\bar{P},\bar{T}$ satisfies (0) $|\bar{P}| = |P|$ (1) $\left|\operatorname{dist}\left(\bigvee_0^u T^i P\right) - \operatorname{dist}\left(\bigvee_0^u \bar{T}^i \bar{P}\right)\right| < \delta$, and (2) $|E(P,T) - E(\bar{P},\bar{T})| < \delta$, then $(\bar{A})$: there is an n and for all m there is a collection C_m of atoms of $\bigvee_{-m}^{-0} \bar{T}^i \bar{P}$, $m(C_m) > 1-\varepsilon$, and if $A \subset C_m$, then $\bar{d}(\{\bar{T}^i \bar{P}\}_1^n, \{\bar{T}^i \bar{P}/A\}_1^n) < \varepsilon$.

<u>Proof</u>. (1) Pick n so that (A) (see definition of V.W.B.) holds for P,T.
(2) Pick δ_1 so that if $|Q| = |\bigvee_1^n T^i P|$, and if $|E(Q/Q_1) - E(Q)| < \delta_1$** $Q \perp^{\varepsilon^2} Q_1$ (we can pick such a δ_1 by Lemma 1, see 4).

(3) Pick ℓ so that $\left|E\left(\bigvee_1^n T^i P / \bigvee_{-\ell}^0 T^i P\right) - nE(P,T)\right| < \frac{1}{2}\delta_1$.

Let $u = n + \ell$.

(4) If δ is small enough, then (3) and the hypotheses ((0), (1), (2)) imply $\left|E\left(\bigvee_1^n \bar{T}^i \bar{P} / \bigvee_{-\ell}^0 \bar{T}^i \bar{P}\right) - nE(\bar{P},\bar{T})\right| < \delta_1$ (the last inequality can be restated as: for all $\bar{\ell} > 0$ $\left|E\left(\bigvee_1^n \bar{T}^i \bar{P} / \bigvee_{-\ell}^0 \bar{T}^i \bar{P}\right) - E\left(\bigvee_1^n \bar{T}^i \bar{P} / \bigvee_{-\ell-\bar{\ell}}^0 \bar{T}^i \bar{P}\right)\right| < \delta_1$).

---

*It was shown recently by Weiss and Ornstein [62] that the V.W.B. condition characterizes the B-processes (that is: if T is a Bernoulli shift and P a finite partition, then P,T is V.W.B.)

**See Appendix 2 for notation.

(5) Pick δ so small that the hypotheses (A), (0), (1) imply there is a collection C_ℓ of atoms of $\bigvee_{-\ell}^{0} \overline{T}^i\overline{P}$ such that $m(C_\ell) > 1 - 2\varepsilon$, and if $A \in C_\ell$, then $\overline{d}(\{\overline{T}^i\overline{P}\}_1^n, \{\overline{T}^i\overline{P}/A\}_1^n) < 2\varepsilon$.

(6) Because of (2) and (4), we have that for all $\overline{\ell} > 0$ there is a collection $\mathscr{C}_{\overline{\ell}}$ of atoms of $\bigvee_{-\ell}^{0} \overline{T}^i\overline{P}$, $m(\mathscr{C}_{\overline{\ell}}) < \varepsilon$, and if $A \subset \mathscr{C}_{\overline{\ell}}$, then $\bigvee_1^n \overline{T}^i\overline{P}$ restricted to A is ε independent of $\bigvee_{-\ell-\overline{\ell}}^{-\ell} \overline{T}^i P$ restricted to A. The lemma follows from (5) and (6).

PROPOSITION 2. If P,T is V.W.B., then P,T is finitely determined.

Proof. (1) We must show that given ε there is a u and a δ such that if (a) $|\overline{P}| = |P|$, (b) $|\text{dist} \bigvee_0^u T^i P - \text{dist} \bigvee_0^u \overline{T}^i\overline{P}| < \delta$, and (c) $|E(P,T) - E(\overline{P},\overline{T})| < \delta$, then $\overline{d}((P,T),(\overline{P},\overline{T})) < \varepsilon$.

(2) Choose u and δ by Lemma 1 so that (a), (b), and (c) imply that there is an n such for all m there is a collection C_m of atoms of $\bigvee_{-m}^{0} \overline{T}^i\overline{P}$, $m(C_m) > 1 - \frac{\varepsilon}{10}$, and if $A \subset C_m$, then $\overline{d}(\{\overline{T}^i\overline{P}\}_1^n, \{\overline{T}^i\overline{P}/A\}_1^n) < \frac{\varepsilon}{10}$. We can assume that (a), (b), and (c) imply $\overline{d}(\{T^iP\}_1^n, \{\overline{T}^i\overline{P}\}_1^n) < \frac{\varepsilon}{10}$.

(3) To show that $\overline{d}((P,T),(\overline{P},\overline{T})) < \varepsilon$, we must, for each k, find sequences $\{P_i\}_1^k$, $\{\overline{P}_i\}_1^k$ of partitions of the same space such that (a) $\text{dist} \bigvee_1^k P_i = \text{dist} \bigvee_1^k T^i P$, $\text{dist} \bigvee_1^K \overline{P}_i = \text{dist} \bigvee_1^K \overline{T}^i\overline{P}$, and (b) $\frac{1}{k}\Sigma|P_i - \overline{P}_i| < \varepsilon$.

Assume we have found the first k partitions. We will now define $\{P_i\}_{k+1}^{k+n}$, $\{\overline{P}_i\}_{k+1}^{k+n}$.

Let A be an atom of $\bigvee_1^k P_i$, $\overline{A}$ an atom of $\bigvee_1^k \overline{P}_i$, and let A_1 and $\overline{A}_1$ be the atoms in $\bigvee_1^k T^i P$ and $\bigvee_1^k \overline{T}^i\overline{P}$ corresponding to A and $\overline{A}$.

We will define $\{P_i\}_{k+1}^{k+n}$ on $A \cap \overline{A}$ so that the distribution of $\bigvee_{k+1}^{k+n} P_i$ on on $A \cap \bar{A}$ is the same distribution of $\bigvee_{k+1}^{k+n} T^i P$ on A. This will insure that $\text{dist} \bigvee_{1}^{k+n} P_i = \text{dist} \bigvee_{1}^{k+1} T^i P$. We will define $\{\bar{P}_i\}_{k+1}^{k+n}$ on $A \cap \bar{A}$ in the analogous way.

Because of (2) we have that there is a collection $\mathscr{C}$ of atoms of $\bigvee_{1}^{k} (P_i \vee \bar{P}_i)$, $m(\mathscr{C}) > 1 - \frac{2\varepsilon}{10}$, and if $A \cap \bar{A} \subset \mathscr{C}$, then

$$\bar{d}\left(\{P_i\}_{k+1}^{k+n}, \left\{P_i/_{A \cap \bar{A}}\right\}_{k+1}^{k+n}\right) < \frac{\varepsilon}{10} \text{ and } \bar{d}\left(\{\bar{P}_i\}_{k+1}^{k+n}, \left\{\bar{P}_i/_{A \cap \bar{A}}\right\}_{k+1}^{k+n}\right) < \frac{\varepsilon}{10}.$$

Thus (by (2)) we have $\bar{d}\left(\left\{\bar{P}_i/_{A \cap \bar{A}}\right\}_{k+1}^{k+n}, \left\{P_i/_{A \cap \bar{A}}\right\}_{k+1}^{k+n}\right) < \frac{3\varepsilon}{10}$.

This last inequality implies that on $A \cap \bar{A}$, $\{P_i\}_{k+1}^{k+n}$ and $\{\bar{P}_i\}_{k+1}^{k+n}$ can be chosen so that $\frac{1}{n} \sum_{i=k+1}^{k+n} \left| P_i/_{A \cap \bar{A}} - \bar{P}_i/_{A \cap \bar{A}} \right| < \frac{3\varepsilon}{10}$. This in turn implies

$$\frac{1}{n} \sum_{i=k+1}^{k+n} \left| P_i - \bar{P}_i \right| < \varepsilon.$$

COROLLARY 3. If P,T is V.W.B., then T acting on $\bigvee_{-\infty}^{\infty} T^i P$ is a Bernoulli shift.

8. Markov processes

In this section we will apply the criterion in section 7 to show that mixing multi-step Markov shifts are Bernoulli. We will then put together various previous results to characterize the B-processes as the closure in the $\bar{d}$-metric of the mixing multi-step Markov processes.

An n-step Markov process (or a multi-step Markov process) on a finite state space is a stationary process with a finite number of possible outputs where the probability of an output given the previous n-outputs is the same as its probability given the previous n+k-outputs for all $k \geq 0$. A 1-step Markov process is called a Markov process.

We say that an n-step Markov process is mixing if the following holds: Let $\{\alpha_i\}_{i=1}^n$ and $\{\beta_i\}_{i=1}^n$ be two n-sequences that are given non-zero probability by our process. Then the probability that $\gamma_{K+i} = \beta_i \; 1 \leq i \leq n$, given that $\gamma_i = \alpha_i \; 1 \leq i \leq n$ is > 0 for all K large enough. (Here γ_i is a sequence of outputs of the process.) It is well known and easy to see that if the above conditional probability is non-zero, then it will tend to the product of the probabilities of $\{\alpha_i\}_1^n$ and $\{\beta_i\}_1^n$ as K tends to infinity.

A (mixing) n-step Markov shift on a finite state space is the transformation arising from a (mixing) n-step Markov process on a finite state space. (That is: the points in the space X are all doubly infinite sequences of outputs of the process. The measure is the unique extension of the measures the process assigns to finite sequences. T is the shift. The partition according to the 0th coordinate of each sequence will be called the Markov generator.)

THEOREM 1.* A mixing n-step Markov shift on a finite state space is isomorphic to a Bernoulli shift.

Proof. The proof will consist of checking the condition V.W.B. (see the beginning of section 7) for the Markov generator P. We will show that given ε there is an M such that if $A \subset \bigvee_{i=0}^{k} T^i P$ for any $k > 0$, then

$$\bar{d}\left(\{T^i P\}_{i=-M}^{i=-1}, \{T^i P/A\}_{i=-M}^{i=-1}\right) < \varepsilon.$$

It will be convenient to check a somewhat stronger property. We will show that given ε there is an M such that $\bigvee_{M}^{M+\ell} T^i P \perp^{\varepsilon} \bigvee_{-k}^{0} T^i P$ for all $\ell > 0$ and $k > 0$. (If we take ℓ very large compared to M it becomes clear that this condition implies V.W.B.)

*This was first proved by Nathaniel Friedman and myself [15].

By the definition of n-step Markov we see that if $k > n$, then $\text{dist}\left(\bigvee_{M}^{M+\ell} T^iP/A\right) = \text{dist}\left(\bigvee_{M}^{M+\ell} T^iP/A'\right)$ where A is any atom in $\bigvee_{-k}^{0} T^iP$ and A' is the atom of $\bigvee_{-n}^{0} T^iP$ containing A. Thus it is enough to show that $\bigvee_{M}^{M+\ell} T^iP \perp^{\varepsilon} \bigvee_{-n}^{0} T^iP$. To do this it will be enough to show that $\bigvee_{-n}^{0} T^iP \perp^{\varepsilon^2} \bigvee_{M}^{M+\ell} T^iP$, and by the same reasoning it will be enough to show that $\bigvee_{-n}^{0} T^iP \perp^{\varepsilon^2} \bigvee_{M}^{M+n} T^iP$. If M is large enough this is insured by the mixing condition.

The multi-step Markov processes play a special role in the theory of stationary processes as is seen from the following lemma.

LEMMA 2. Any stationary process can be approximated arbitrarily well by multi-step Markov processes in the vague topology. That is, given a process P,T and an n we can find an n-step Markov process P_n,T_n that gives the same measure to sequences of length n as does P,T. If T is mixing, then T_n will also be mixing. Also, $E(P_n,T_n) \geq E(P,T)$.

Proof. We construct P_n,T_n so that the distribution of the last n - 1 letters printed out is the same as the distribution of (n - 1)-sequences under P,T. The probability of the next letter will be such that the distribution of the last (n - 1) letters together with the letter being printed is the same as the distribution of n-sequences under P,T. (It is obvious that this can be done—merely copy P,T.) This can be continued because to print out a letter the next time we again look at the last (n - 1) letters printed out. Their distribution is the same as the distribution of the last (n - 1) letters in the previous step (because P,T was stationary). It is thus obvious that this defines a stationary process which is an n-step Markov process, and which gives n-sequences the same measure as does P,T.

We should point out that we have given a canonical description of P_n,T_n.

P_n, T_n can be thought of, roughly, as the process P,T, modified by erasing its memory of what it did more than n-steps in the past.

It is obvious, furthermore, that $E(P_n, T_n) \geq E(P,T)$.

If T is mixing, then so is T_n. We see this as follows: It is easy to see (and well known) that T_n will be mixing if there is a K such that: Let $\{\alpha_i\}_1^n$ and $\{\beta_i\}_1^n$ be two n-sequences that have probability > 0 under P_n, T_n. Then the probability that $\gamma_{K+i} = \beta_i$, $1 \leq i \leq n$, given that $\gamma_i = \alpha_i$, $1 \leq i \leq n$ is > 0. The above statement is true for the process P,T (instead of P_n, T_n) because T is mixing. This will imply that it holds for P_n, T_n because any finite sequence that has non-zero probability under P,T also has non-zero probability under P_n, T_n (this is obvious by induction).

Let us collect here some of the facts we have already proven about stationary processes (see section 6).

THEOREM 3. The class of B-processes is closed in the $\bar{d}$-metric.

COROLLARY 4. The B-processes are exactly those processes that can be approximated arbitrarily well in the $\bar{d}$-metric by finite codings of roulette wheels.

THEOREM 5.* The B-processes are exactly those processes that are finitely determined.

THEOREM 6. The B-processes are the closure in the $\bar{d}$-metric of the mixing multi-step Markov processes.

Proof. This theorem follows easily from the fact that mixing n-step Markov shifts are Bernoulli, and from Theorems 1, 3, and 5. We can find a sequence of n-step Markov processes that converges to any process in entropy and in the vague topology—if the process is a B-process, then we get convergence in the $\bar{d}$-metric (Theorem 5). On the other hand, mixing n-step Markov processes are B-process (Theorem 1), and hence (Theorem 3) the closure of these processes in the $\bar{d}$-metric contains only B-processes.

*Theorem 5′. The B-processes are exactly those processes that are V.W.B. (see footnote on p. 42).

Remark. Theorems 5′ and 6 can be interpreted as saying that the B-processes are exactly those processes where the influence of the sufficiently distant past becomes negligible. (Theorem 5′ says that if P,T is not a B-process, then it is not V.W.B., and hence we can predict essentially nothing about the long term future from the past. Theorem 6 says that it is some distance α away from any process with a finite memory.)

9. Infinite entropy, generalized Bernoulli shifts, and another criterion for Bernoulli shifts

In this section we will generalize the definition of Bernoulli shift to cover the case of infinite entropy and show that two generalized Bernoulli shifts of the same (possibly infinite) entropy are isomorphic.* We will show that the theorem on factors holds for generalized Bernoulli shifts. We will also get another criterion for Bernoulli shifts.

DEFINITION. We define a generalized Bernoulli shift as follows: Let Y be a measure space of total measure 1 consisting of a part (possibly empty) that is isomorphic to the unit interval and a part (possibly empty) consisting of a finite or countable number of atoms. Let $X = \prod_{-\infty}^{\infty} Y_i$ be the product of a countable number of copies of Y with the product measure. Let T be the shift on X (each point in X is a sequence $\{y_i\}$ of points in Y. $T\{y_i\} = \{y'_i\} = \{y'_i\}$ where $y'_i = y_{i+1}$).

If the part of Y which is isomorphic to the unit interval has non-zero measure, then the entropy of X is infinite. Otherwise X is atomic, and the entropy of T will be $-\Sigma\, p_i \log p_i$ where p_i is the measure of the ith atom of Y.

LEMMA 1. Let T be a Bernoulli shift of finite entropy, and P a finite partition such that $E(P,T) < E(T)$, and T^iP are independent. Then given ε there are P' and R' such that (1) $P',T \sim P,T$, (2) $|P - P'| < \varepsilon$,

*The case of infinite partitions of finite entropy was first worked out by Smorodinsky [18].

(3) $P' \vee R'$ generates, and (4) the σ-algebras $\bigvee_{-\infty}^{\infty} T^i P'$ and $\bigvee_{-\infty}^{\infty} T^i R'$ are independent, and $T^i R'$ are independent.

Remark. R' can be chosen to have any distribution whose entropy equals $E(T) - E(P,T)$. This follows from the isomorphism theorem for Bernoulli shifts.

Proof. (1) Let $\alpha = E(T) - E(P,T)$. Let $\overline{T}$ be a Bernoulli shift with independent generator $\overline{P} \vee \overline{R}$ where $\overline{P} \perp \overline{R}$ and dist P = dist $\overline{P}$ and $E(\overline{R}) = \alpha$ (then $E(\overline{T}) = E(T)$).

(2) Because of Corollary 3, and Propositions 9 and 11 of section 4, there is a δ such that if R is a partition satisfying (a) $|R| = |\overline{R}|$, (b) $|E(R \vee P,T) - E(\overline{R} \vee \overline{P},\overline{T})| < \delta$, and (c) $|\text{dist } R \vee P - \text{dist } \overline{R} \vee \overline{P}| < \delta$, then there are partitions R_1 and P_1 such that $|R_1 \vee P_1 - R \vee P| < \frac{1}{2}\varepsilon$, $R_1 \vee P_1, T \sim \overline{R} \vee \overline{P}, \overline{T}$, and $R_1 \vee P_1$ generate.

(3) There is a $\delta_1 < \delta$ such that if (b_1) $|E(R \vee P,T) - E(\overline{R} \vee \overline{P},\overline{T})| < \delta_1$ and (c_1) $|\text{dist } R - \text{dist } \overline{R}| < \delta_1$, then (c) $|\text{dist } R \vee P - \text{dist } \overline{R} \vee \overline{P}| < 2\delta$. We can see this as follows: Using (b_1) we get $E(R \vee P) \geq E(R \vee P,T) \geq E(T) - \delta_1 \geq E(P) + E(\overline{R}) - \delta_1$. ($c_1$) implies that if δ_1 is small enough, $E(R \vee P)$ is so close to $E(P) + E(R)$ that R must be $\frac{1}{10}\delta$ independent of P (Lemma 1, section 4). This and (c_1) then imply (c).

The rest of the proof will be devoted to finding an R satisfying (a), (b_1), and (c_1).

(4) Let Q be a refinement of P such that $E(T) - \frac{1}{100}\delta_1 < E(Q,T) < E(T)$. Let $\beta = E(T) - E(Q,T)$.

(5) Pick $\gamma < \delta_1$ so that if $|Q_1 - Q| < \gamma$, then $|E(Q_1,T) - E(Q,T)| < \frac{1}{100}\delta_1$.

(6) Call an atom of $\bigvee_{-n+1}^{0} T^i Q$ a good Q-n-atom if its measure is between $\frac{1}{2}^{(E(Q,T) \pm \frac{1}{100}\beta)n}$. Call an atom of $\bigvee_{-n+1}^{0} T^i P$ a good P-n-atom if its measure is between $\frac{1}{2}^{(E(P,T) + \frac{1}{100}\beta)n}$. Call an atom of $\bigvee_{-n+1}^{0} \overline{T}^i \overline{R}$ a good

$\overline{R}$-n-atom if its measure is between $2^{-(E(\overline{R},\overline{T}) \pm \frac{1}{100}\beta)n}$, and if the distribution of its $\overline{R}$-n-name is within $\frac{1}{100}\delta_1$ of the distribution of $\overline{R}$. (That is: each atom of $\bigvee_{-n+1}^{0} \overline{T}^i\overline{R} = \bigcap_{-n+1}^{0} \overline{T}^i\overline{R}f^{(i)}$ when the $\overline{R}f^{(i)}$ are atoms of $\overline{R}$. We are requiring that the number of $f^{(i)}$ that equal j divided by n be close to the measure of $\overline{R}^j$.)

(7) Call a good P-n-atom very good if more than $1 - \frac{1}{100}\gamma$ of it is covered by good Q-n-atoms. Call the good Q-n-atoms in a very good P-n-atom very good. The Shannon-McMillan-Brieman theorem implies that if n is large enough, then (a) all but $1 - \frac{1}{100}\gamma$ of the Q-n-atoms are very good. It also implies that (b) each very good P-n-atom contains fewer very good Q-n-atoms than there are good $\overline{R}$ atoms. (We see (b) as follows: If n is large enough, the number of good $\overline{R}$-n-atoms $> 2^{(E(\overline{R},\overline{T}) - \frac{1}{100}\beta)n} = 2^{[E(T) - E(P,T) - \frac{1}{100}\beta]n}$. The number of very good Q-n-atoms in a very good P-n-atom is $< 2^{[E(Q,T) - E(P,T) + \frac{2}{100}\beta]n}$.)

(8) We can (because of 7(b)) assign to each very good Q-n-atom in a fixed very good P-n-atom a good $\overline{R}$-n-atom (so that no 2, Q-n-atoms in the same P-n-atom are assigned the same $\overline{R}$-n-atom).

(9) Using the R-K theorem we can find a set F such that T^iF, $0 \leq i \leq n-1$ are disjoint and $m\left(\bigcup_{i=0}^{n-1} T^iF\right) \geq 1 - \frac{1}{100}\gamma$, and such that

$$\operatorname{dist}\left(\bigvee_{-n+1}^{0} T^iQ/F\right) = \operatorname{dist} \bigvee_{-n+1}^{0} T^iQ.$$

Call the part of a Q-n-atom that lies in F a Q-n-F-atom. Make the same definition for good and very good Q-n-atoms and P-n-atoms.

(10) We will now define R as follows: Let A be a very good Q-n-F-atom. Let $\bigcap_{-n+1}^{0} \overline{T}^i\overline{R}f^{(i)}$ be the good $\overline{R}$-n-atom assigned to A. Then T^iA will be in

$Rf^{(i)}$. (Thus the R-n-name of points in A will be the same as the $\overline{R}$-n-name of points in the $\overline{R}$-n-atoms assigned to A.)

Let F_1 be the union of the very good Q-n-F-atoms. We have defined R on $\bigcup_0^{n-1} T^i F_1$. We can define R in an arbitrary way on the rest of the space. Because of 7(a) we have that (a) $m(F_1) > (1 - \frac{1}{100}\gamma)\, m(F)$.

(11) R satisfies 2(a) and 3(c_1) (i.e., has good distribution) because the distribution of the R-n-name of each point in F_1 will itself have a distribution within $\frac{1}{100}\delta_1$ of the dist $\overline{R}$. (See (6) for the definition of good-$\overline{R}$-n-atoms and (10) for the relation to the R-n-name.) This shows that

$|\text{dist } R - \text{dist } \overline{R}| < \delta_1$, because $m\left(\bigcup_0^{n-1} T^i F\right) > 1 - \frac{1}{100}\gamma$ and $m(F_1) > (1 - \frac{1}{100}\gamma)\, m(F)$.

(12) $R \vee P$ satisfies 3(b_1) (has good entropy) because $\bigvee_{-n}^{n} T^i(R \vee P \vee F)$ refines Q on $\bigcup_0^{n-1} T^i F_1$ (here F denotes the partition consisting of F and its complement). Thus $\bigvee_{-n}^{n} T^i(R \vee P \vee F) \overset{\gamma}{\supset} Q$, and by the choice of γ $E(R \vee P \vee F, T) > E(Q,T) - \frac{1}{100}\delta_1 \geq E(T) - \frac{2}{100}\delta_1$ (by (4)). Because $E(F) < \frac{1}{100}\delta_1$, we get that $E(R \vee P, T) \geq E(T) - \delta_1 = E(\overline{R} \vee \overline{P}, \overline{T}) - \delta_1$, and hence using (11) and the independence of $\overline{T}^i(\overline{R} \vee \overline{P})$, $|E(R \vee P, T) - E(\overline{R} \vee \overline{P}, \overline{T})| - \delta_1$.

(13) Because of (2) and (3) we can find R' and P' such that (a) $R' \vee P'$, $T \sim \overline{R} \vee \overline{P}, \overline{T}$, (b) $|R' \vee P' - R \vee P| < \frac{1}{2}\varepsilon$, and (c) $R' \vee P'$ generate. (a), (b), and (c), and the properties of $\overline{P}$, $\overline{R}$, and $\overline{T}$ imply that P' and R' satisfy the conclusions of the lemma.

DEFINITION. We say that T is an increasing union of Bernoulli shifts if there is an increasing sequence of invariant σ-algebras σ_i such that T restricted to each σ_i is a Bernoulli shift of finite entropy, and the smallest σ-algebra containing all the σ_i is the entire σ-algebra.

LEMMA 2. Let T and $\overline{T}$ be increasing unions of Bernoulli shifts with invariant σ-algebras σ_i and $\overline{\sigma}_i$ respectively. We will assume that the entropy of T restricted to σ_i is the same as the entropy of $\overline{T}$ restricted to $\overline{\sigma}_i$. Then T and $\overline{T}$ are isomorphic (in the measure algebra sense, and if T and $\overline{T}$ act on Lebesgue space and $\cup\sigma_i$ and $\cup\overline{\sigma}_i$ separate points, then T and $\overline{T}$ are isomorphic in the pointwise sense).

Proof. Because of the isomorphism theorem for Bernoulli shifts we can pick ${}^1_1P$ and ${}^1_1\overline{P}$ such that ${}^1_1P, T \sim {}^1_1\overline{P}, \overline{T}$, ${}^1_1P$ generates σ_1, and ${}^1_1\overline{P}$ generates $\overline{\sigma}_1$.

We now apply the previous lemma to σ_2 and $\overline{\sigma}_2$ to get: given ε_1 there exist ${}^1_2P, {}^2_2P$ and ${}^1_2\overline{P}, {}^2_2\overline{P}$ satisfying (1) $|{}^1_2P - {}^1_1P| < \varepsilon_1$ and $|{}^1_2\overline{P} - {}^1_1\overline{P}| < \varepsilon$. (2) ${}^1_2P \vee {}^2_2P$ generates σ_2 and ${}^1_2\overline{P} \vee {}^2_2\overline{P}$ generates $\overline{\sigma}_2$.

We will also have : $\bigvee_{-\infty}^{\infty} T^i({}^1_2P) \perp \bigvee_{-\infty}^{\infty} T^i({}^2_2P)$, $\bigvee_{-\infty}^{\infty} \overline{T}^i({}^1_2\overline{P}) \perp \bigvee_{-\infty}^{\infty} \overline{T}^i({}^2_2\overline{P})$ and ${}^1_2P, T \sim {}^1_2\overline{P}, \overline{T} \sim {}^1_1P, T \sim {}^1_1\overline{P}, \overline{T}$ and ${}^2_2P, T \sim {}^2_2\overline{P}, \overline{T}$. Thus we get

(3) $\left({}^1_2P \vee {}^2_2P\right), T \sim \left({}^1_2\overline{P} \vee {}^2_2\overline{P}\right), \overline{T}$.

If we repeat this process once more we get ${}^1_3P, {}^2_3P, {}^3_3P$ and ${}^1_3\overline{P}, {}^2_3\overline{P}, {}^3_3\overline{P}$ satisfying $(\overline{1})$ $|{}^1_3P \vee {}^2_3P - {}^1_2P \vee {}^2_2P| < \varepsilon_2$, $(\overline{2})$ $\left(\bigvee_{i=1}^{3} {}^i_3P\right)$ generate σ_3 and $\left(\bigvee_{i=1}^{3} {}^i_3\overline{P}\right)$ generate $\overline{\sigma}_3$, and $(\overline{3})$ $\left(\bigvee_{i=1}^{3} {}^i_3P\right), T \sim \left(\bigvee_{i=1}^{3} {}^i_3\overline{P}\right), \overline{T}$.

Continuing in this way we get ${}^i_nP$ and ${}^i_n\overline{P}$ $1 \le i \le n$ satisfying

$(\hat{1})$ $\left|\left(\bigvee_{i=1}^{n-1} {}^i_{n-1}P\right) - \left(\bigvee_{i=1}^{n-1} {}^i_nP\right)\right| < \varepsilon_{n-1}$, $\left|\left(\bigvee_{i=1}^{n-1} {}^i_{n-1}\overline{P}\right) - \left(\bigvee_{i=1}^{n-1} {}^i_n\overline{P}\right)\right| < \varepsilon_{n-1}$, $(\hat{2})$ $\bigvee_{i=1}^{n} {}^i_nP$ generate σ_n and $\bigvee_{i=1}^{n} {}^i_n\overline{P}$ generate $\overline{\sigma}_n$, and

$(\hat{3})$ $\left(\bigvee_{i=1}^{n} {}^i_nP\right), T \sim \left(\bigvee_{i=1}^{n} {}^i_n\overline{P}\right), \overline{T}$.

If the ε_i are chosen small enough, ${}_n^iP$ converge (as $n \to \infty$, i fixed) to ${}^iP$, and it is clear that (4) $\left(\bigvee_1^\infty {}_iP\right), T \sim \left(\bigvee_1^\infty {}_i\overline{P}\right), \overline{T}$.

We will be finished if we show that $\bigvee_1^\infty {}_iP$ and $\bigvee_1^\infty {}_i\overline{P}$ generate.

We will do this as follows: Let Q_n be a generator for σ_n such that Q_{n+1} refines Q_n. Thus $\bigvee_{-\infty}^{\infty} T^i\left(\bigvee_{j=1}^{n} {}_n^jP\right) \supset Q_n$ and for some K_n

$\bigvee_{-K_n}^{K_n} T^i\left(\bigvee_{j=1}^{n} {}_n^jP\right) \overset{\frac{1}{2}^n}{\supset} Q_n$. If we choose ε_n, ε_{n+1}, etc. properly, we will get that $\bigvee_{-K_n}^{K_n} T^i\left(\bigvee_{j=1}^{n} {}^jP\right) \overset{\frac{1}{2}^{n-1}}{\supset} Q_n$. This shows that $\bigvee_1^\infty {}_iP$ generates. The same argument works for $\overline{T}$.

THEOREM 3. All generalized Bernoulli shifts of infinite entropy are isomorphic.

THEOREM 4. If $E(T) = \infty$ and T is an increasing union of Bernoulli shifts, then T is a generalized Bernoulli shift.

Proof of Theorems 3 and 4. Let T be defined as follows. Let X be the product of copies Y_i, $-\infty < i < \infty$ of the unit interval, with product measure, and let T be the shift.

T is a generalized Bernoulli shift of infinite entropy, and if r_i is a monotone increasing unbounded sequence we can find $\sigma_{i+1} > \sigma_i$ such that the $\bigcup \sigma_i$ generate, and T restricted to σ_i is a Bernoulli shift of entropy r_i. This fact and the previous lemma prove Theorem 2.

Theorem 1 follows from Theorem 2 because any generalized Bernoulli shift of infinite entropy is an increasing union of Bernoulli shifts.

THEOREM 5. If $E(T) < \infty$ and T is an increasing union of Bernoulli shifts (or if T is a generalized Bernoulli shift);* then T is a Bernoulli shift (i.e., has a finite independent generator).

Remark. The above theorem allows us to prove that a transformation is Bernoulli without first finding a generator.

Proof. The theorem follows from our previous lemma and the following: (1) If T is a Bernoulli shift and if r_i are numbers such that $r_{i+1} > r_i$ and $\sup r_i = E(T)$, then we can find invariant σ-algebra σ_i such that T restricted to σ_i is a Bernoulli shift of entropy r_i.

We will get (1) immediately from the theorem that factors of Bernoulli shifts are Bernoulli shifts and from: (2) Let T be a Bernoulli shift and let σ_1 be an invariant σ-algebra such that the entropy of T restricted to σ_1 (denoted by $E(\sigma_1)$) is strictly less than E(T), then given $r, E(\sigma_1) < r < E(T)$ we can find an invariant σ-algebra σ_2 such that $\sigma_2 \supset \sigma_1$ and $E(\sigma_2) = r$. We get (2) as follows: Let P_1 be a generator for T acting on σ_1 and P_2 a generator for T. Deform P_2 continuously, through P^t, into a partition with only one atom. Then $E(P_1 \vee P^t, T)$ will go continuously from E(T) to $E(\sigma_1)$.

THEOREM 6. A factor of a generalized Bernoulli shift of infinite entropy is a generalized Bernoulli shift.

Proof. It is enough to show that (1) if P is a finite partition (and T a generalized Bernoulli shift), then T acting on $\bigvee_{-\infty}^{\infty} T^i P$ is Bernoulli. (This is so because any factor has a countable generator and would therefore be an increasing union of Bernoulli shifts.) (2) If we represent our generalized Bernoulli shift of infinite entropy as the shift on the product of copies of the unit interval, it is obvious that given ε_i there is an increasing sequence σ_i of invariant σ-algebras such that T restricted to σ_i is a Bernoulli shift

*If we worked with countable partitions, as does Smorodinsky [18], in the isomorphism theorem, there would be no distinction between Bernoulli shifts of finite entropy and generalized Bernoulli shifts of finite entropy. This is in a sense more natural, but I think that it is sufficiently easier to work with finite partitions to warrant the complication arising here.

of finite entropy and $\sigma_i \overset{\varepsilon_i}{\supset} P$. (3) Now let ${}^iP$ be a finite partition in σ_i such that $|{}^iP - P| < 2\varepsilon_i$. Since ${}^iP$ is a B-process, (3) implies that P,T can be approximated arbitrarily well in the $\bar{d}$-metric by B-processes. Hence, by Theorem 1, section 6, T acting on $\bigvee_{-\infty}^{\infty} T^iP$ is Bernoulli.

PART II

Flows

We will call a flow S_t a Bernoulli flow if S_1 is a Bernoulli shift. (We will prove later that if S_t is a Bernoulli flow, then S_{t_0} for each fixed t_0 is a Bernoulli shift.)

The next three sections will deal with flows. The main results are the following: We will first show that a certain simple flow is Bernoulli (thus proving the existence of Bernoulli flows). We will then get an isomorphism theorem for Bernoulli flows of finite entropy (i.e., if S_t and $\overline{S}_t$ are Bernoulli flows of finite entropy, and if we normalize so that $E(S_1) = E(\overline{S}_1)$, then S_t is isomorphic to $\overline{S}_t$). In the process we will show that if S_1 is Bernoulli, then S_{t_0} is Bernoulli, for all t_0, and we will get the analog of Sinai's theorem which will tell us that any flow of non-zero entropy has a Bernoulli factor. We will then extend the isomorphism theorem to Bernoulli flows of infinite entropy and give some examples of these.

10. A Bernoulli flow

We will now apply the criterion of V.W.B. (see section 7) to show that a certain flow S_t has the property that for each fixed t, S_t is a Bernoulli shift of finite entropy.

<u>Definition of S_t</u>. The flow, S_t, which was previously shown to be K by Totoki [27], is described as follows: Let T be the 2-shift acting on Y and let $Q = \{Q_0, Q_1\}$ be the independent generator. Let f be a function on Y that takes on the value α on Q_0 and β on Q_1. Assume $\frac{\alpha}{\beta}$ is irrational and $\frac{1}{2}(\alpha + \beta) = 1$. X will be the part of $Y \times R$ that lies under the graph of f (i.e., t will increase with unit speed until it hits the graph of f (until $t = f(y)$)). It then goes to $(T(y), 0)$ and continues to move up at unit speed.

Let $P = (P_0, P_1)$ be the partition of X defined as follows: P_0 is the set of $(y,t), y \in Q_0, 0 \leq t \leq \alpha$ and P_1 the set of $(y,t), y \in Q_1, 0 \leq t \leq \beta$.

It will be shown that P generates under some S_t and therefore S_t has finite entropy.

If we consider the process defined by S_t,P then we see that S_t is in some sense the natural generalization of the Bernoulli shifts. The process S_tP will be described in more detail below.

DEFINITION. By the continuous P-name of x we will mean the line R partitioned in such a way that t in R is in the ith atom of this partition if $S_t(x)$ is in the ith atom of P.

By the continuous P-name of x of length n we will mean (0,n) partitioned as above.

The flow S_t together with the partition P give us a continuous stationary process as follows: Each point x in X has a continuous P-name, and X gives us a probability measure on these names. We will call this the X process. Totoki showed that this flow, first studied by Kakutani, is K [27]. If we look only at the continuous P-name of the points in Y with the measure given by the measure on Y, we get another process (not stationary) which we will call the Y process.

The Y process can be described probabilistically as follows: At time 0 we pick a 0 with probability $\frac{1}{2}$ and stay in the 0 state for time α, or pick a 1 with probability $\frac{1}{2}$ and stay in the 1 state for time β. We then pick a 0 or 1 with probability $\frac{1}{2}$ (independently of what happened before) and stay in the 0 state for time α and the 1 state for time β, etc. This is equivalent to doing a random walk where we jump α or β with probability $\frac{1}{2}$ and label the points we jumped over with an α jump 0, and those we jumped over with a β jump 1.

The X process differs from the Y process in that, instead of starting at 0, the starting time has a probability distribution u where $u = u' + u''$, and u' has total measure $\frac{1}{2}\alpha$ and is uniform on $0 \leq t \leq \alpha$, and u'' is uniform on $0 \leq t \leq \beta$.

Let B be an atom in $\bigvee_{-N}^{0} T^i Q$. Let $A = \bigcup_{t_0 \leq t \leq t_1} S_t B$ where $0 \leq t_0$ and t_1 is $\leq \alpha$ if $B \subset Q_0$ and $\leq \beta$ if $B \subset Q_1$. If we normalize A to have measure 1, A will give us a probability measure on the continuous P-names of the points y in A. This gives us a process which we will call the A process. It is easy to see that the A process differs from the X (or Y) process, only in the distribution of the starting time (the starting time has uniform distribution on $\beta - t$ or $\alpha - t$, $t_0 \leq t \leq t_1$, depending on whether B is in Q_0 or Q_1).

The process P, S_γ is obtained from the Y process by looking only at t of the form $k\gamma$ (where k is an integer). Similarly, the distribution of P-names under S_γ restricted to points in A is obtained by looking at the A process at times of the form $k\gamma$.

LEMMA 1. If γ is small enough then P generates under S_γ.

Proof. It is enough to show that, except for a set of measure 0, the P-names under S_γ separate points.

It is easy to see that the continuous P-names separate points. (This is because the Q-names under T separate points in Y.)

It is also easy to see that the P-name of x under S_γ determines the continuous name of x except for a possible translate of $\leq \gamma$.

Except for a set of x of measure 0, the continuous P-name of x will have a collection of t_i such that at t_i we switch from 0 to 1, and the t_i are dense mod γ. (This is so because, except for a set of measure 0, each finite sequence of 0's and 1's must occur.)

The last fact implies that 2 different translates of the continuous P-names of x will label some point of the form $k\gamma$ differently.

THEOREM 2. S_t is a Bernoulli flow.

Proof. We must now examine the partition $\bigvee_{-N}^{0} S_\gamma^i P$. It will be convenient to consider the partition R_N which is a refinement of $\bigvee_{-N}^{0} S_\gamma^i P$. R_N will consist of atoms A, each one of which has the form: B will be some atom in $\bigvee_{-N}^{0} T^i Q$. $A = \bigcup_{t_1 \le t \le t_0} S_t B$ where t_0 and t_1 depend on A and $0 \le t_1$ and $t_0 \le \alpha$ if B is in Q_0 and $\le \beta$ if B is in Q_1.

We will show that given ε there is an n such that for all N

$$\bar{d}\left(\{S_j^i P\}_1^n, \{S_j^i P/A\}_1^n\right) < \varepsilon, \text{ where } A \in R_N.$$

Since a sequence of length n of ordered partitions of a probability space is equivalent to putting a probability measure on sequences of length n, we can state what we must show as follows: The S_γ names of length n with the measure induced by the X process is within ε in the $\bar{d}$ sense of the S_γ names of length n with the measure induced by the A process.

We must thus show that given ε there is an n such that if we start the Y process with two different starting distributions u_1 and u_2 whose supports are in an interval of length $< \max(\alpha, \beta)$, then we have (A): The measures induced on the sequences $k\gamma$ $1 \le k \le n$, (i.e., the S_γ names of length n) are within ε in the $\bar{d}$ sense.

Let us first note that (A) holds for all n large enough if the support of u_1 and u_2 are both in an interval whose length is small compared to ε.

We prove the above statement as follows: The renewal theorem implies that given δ there is an M such that if J is an interval of length $< \delta$ whose left end point is $>$ than M, then the probability that the X process changes from 0 to 1 or 1 to 0 in J is less than 2δ. This implies that given ε there is an N such that with probability $> 1 - \frac{1}{2}\varepsilon$, the number of k, $1 \le k \le n$, $n > N$ such that the process changes from 0 to 1 or 1 to 0 between $k\gamma - \frac{1}{10}\varepsilon^2$ and $k\gamma + \frac{1}{10}\varepsilon^2$ is $< \frac{1}{2}\varepsilon n$. This means that if we change the starting distribution of the X process by $< \frac{1}{10}\varepsilon^2$, then with probability $> 1 - \frac{1}{2}\varepsilon$ we will change the process at times $k\gamma$ for $< \frac{1}{2}\varepsilon n$ of the $k(1 \le k \le n)$. This implies that if u_1 and u_2 both have support in an interval of length $< \frac{1}{10}\varepsilon^2$, then the measure induced by the u_1 and u_2 processes on sequences $k\gamma$, $1 \le k \le n$ (i.e., the S_γ names of length n) will differ by $< \varepsilon$ in the $\bar{d}$ sense.

We will now remove the $\frac{1}{10}\varepsilon^2$ restriction on the supports of u_1 and u_2 as follows: Let I be an interval of length α (assume $\alpha > \beta$). Partition I into disjoint interval J_i, $1 \leq i \leq m$ of the same length and smaller than $\frac{1}{10}\varepsilon^2$. The renewal theorem implies that if we translate I (and the J_i) by more than M, then the difference between the probabilities that the random walk that starts with distributions u_1 and u_2 hits I for the first time in J_i is $< \frac{1}{10}\varepsilon \cdot \frac{1}{m}$. (The random walk we refer to is, of course, the one that jumps α or β with probability $\frac{1}{2}$.) Note that M does not depend on u_1 or u_2 so long as they both have support in a fixed interval.

Thus the u_1 and u_2 process will induce measures on sequences $k\gamma$, $r < k \leq r + n$, where $r\gamma > M$, that are close in the $\bar{d}$ sense, if n is large compound to r we get the desired result.

Remark. Exactly the same proof will show that a whole class of flows are Bernoulli. The class is described as follows: Let T be a Bernoulli shift acting on Y and let Q be its independent generator. Let f be a function, measurable with respect to Q, and whose range does not lie in an arithmetic progression containing 0. Let S_t be the flow built under the function f. Then S_t is a Bernoulli flow.

Note that the condition on the range of f is exactly what we need for the renewal theorem.

If T is a generalized Bernoulli shift of infinite entropy and f a function taking on a finite number of values and measurable with respect to the independent generator of T (and whose range does not lie in an arithmetic progression containing 0), then the above argument shows that there is an increasing sequence of invariant σ-algebras which span the entire σ-algebra and such that S_t restricted to σ_i is a Bernoulli flow of finite entropy.

More examples: A continuous time Markov process on a finite state space gives rise to a Bernoulli flow of infinite entropy. This type of example was discovered by J. Feldman and M. Smorodinsky [22].

Here is a rough outline of how this goes: From the Markov process we get a measure on paths (the paths being a partition of the reals according to which atom of the state space you are in at time t). S_t shifts each path by t.

Let P be the partition of the paths according to which state it is in at t = 0. Let $P_n = \bigvee_{i=1}^{n} S^i_{\frac{1}{n}} P$. Then $\bigvee_{-\infty}^{\infty} S^i_1 P_n = \sigma_n$ form an increasing sequence of σ-algebras, invariant under S_1, and S_1 acting on σ_n is Bernoulli because it is a Markov shift on σ_n (p. 44). Since the σ_n generate the whole σ-algebra we get that S_1 is Bernoulli (pp. 35, 36).

We can see that the entropy is infinite as follows: If the entropy were finite, then $E(S_1) = \alpha$, $E(S_{\frac{1}{n}}) = \frac{\alpha}{n}$, and $E(P,S_{\frac{1}{n}}) \leq \frac{\alpha}{n}$. However, since at least one of the transition probabilities must have a non-zero derivative at 0, we get that for some i and j and $\beta > 0$, the probability of going from i to j in time $\frac{1}{n}$ is $> \frac{\beta}{n}$. Thus, $E(P,S_{\frac{1}{n}})$ would be greater than a constant times $-\frac{\beta}{n} \log \frac{\beta}{n}$.

Another example of a Bernoulli flow of infinite entropy is provided by Brownian motion confined to a rectangular box (when the paths reflect off the walls). This example is shown to be Bernoulli in joint work with P. C. Shields [55, 24].

11. The isomorphism theorem for Bernoulli flows of finite entropy

Part I (preliminaries)

DEFINITIONS. A flow is a family S_t of invertible measure-preserving transformations of a measure space X* such that $S_{t_1}\left(S_{t_2}(x)\right) = S_{t_1+t_2}(x)$ for all $x \in X$. (In what follows we will always assume m(X) = 1.)

The flow is said to be measurable if for each measurable set $E \subset X$ the set of all (t,x) such that $S_t(x) \in E$ is a measurable set in the product of the real line and X.

The flows S_t acting on X and $\overline{S}_t$ acting on $\overline{X}$ are said to be isomorphic if

*We will not assume that the σ-algebra of X separates points.

we can find subsets $X_1 \subset X$ and $\overline{X}_1 \subset \overline{X}$ such that $m(X - X_1) = 0$ and $m(\overline{X} - \overline{X}_1) = 0$, X_1 and $\overline{X}_1$ are invariant (i.e., if $x \in X_1$ then $S_t(x) \in X_1$ for all t), and there exists a measure-preserving invertible φ mapping X_1 onto $\overline{X}_1$ such that $\overline{S}_t(\varphi(x)) = \varphi(S_t(x))$ for all $x \in X_1$ and all t.

A flow is said to be Bernoulli if S_t is a Bernoulli shift for some t.

By a flow built under a function we mean the following: We start with a measure space Y and a transformation T on Y. Let g be an integrable non-negative function on Y. The space X on which the flow acts is the part of $Y \times R$ that lies below the graph of g. Each point (y,r) in X moves straight up at unit speed until it hits y,g(y). It then goes to (Ty,0) and continues to move up at unit speed. ($S_t(y,r) = (y,r+t)$ if $r + t < g(y)$. $S_t(y,r) = (Ty, r + t - g(y))$ if $g(y) < r + t$ and $r + t - g(y) < g(Ty)$, etc.)

We will refer to Y as the "base" of the flow when represented in this way.

We will, as always, assume our flows to be ergodic. (That is, the only measurable sets that are invariant under the flow have measure 0 or 1.)

There is a theorem due to Ambrose [19] and extended by Ambrose and Kakutani which says that every ergodic measurable flow is isomorphic to a flow built under a function. We will therefore always assume that the flows that we are dealing with are flows built under a function.

The purpose of section 11 is to prove the following theorem:

THEOREM 0. Let S_t and $\overline{S}_t$ be Bernoulli flows (acting on X and $\overline{X}$) such that $E(S_1) = E(\overline{S}_1) < \infty$. We assume also that X and $\overline{X}$ are Lebesgue spaces whose σ-algebras separate points. Then S_1 and $\overline{S}_1$ are isomorphic.

Remark. If we drop the assumption that X and $\overline{X}$ are Lebesgue spaces whose σ-algebra separate points, then we still get an isomorphism in the measure algebra sense.

We will first extend the definition of the $\bar{d}$-metric to continuous time.

DEFINITION. Let S_t and $\overline{S}_t$ be flows acting on X and $\overline{X}$, and let P and $\overline{P}$ be partitions of X and $\overline{X}$, $d(\{S_tP\}_a^b, \{\overline{S}_t\overline{P}\}_a^b)$ will be defined as follows: Let ψ be an invertible measure-preserving map of X onto $\overline{X}$. Let

$\bar{d}_\psi(\{S_tP\}_a^b, \{\bar{S}_t\bar{P}\}_a^b) = \frac{1}{|b-a|}\int_a^b |\bar{S}_t\bar{P} - \psi(S_tP)|\,dt$; $\bar{d}$ will be the inf of $\bar{d}_\psi$ over all ψ.

$\underline{\bar{d}((\bar{P},\bar{S}_t),(P,S_t))}$ will denote $\sup_{a,b} \bar{d}(\{\bar{S}_t\bar{P}\}_a^b, \{S_tP\}_a^b$. If $\bar{d}((P,S_t),(\bar{P},\bar{S}_t)) = 0$, we will write $\underline{P,S_t \sim \bar{P},\bar{S}_t}$.

Remark. It is easy to see that $\bar{d}_\psi(\{\bar{S}_t\bar{P}\}_a^b, \{S_tP\}_a^b) = \frac{1}{|a-b|}\int_{\bar{X}}\left[\int_a^b f\left[\bar{S}_t(\bar{x}), S_t(\psi^{-1}(\bar{x}))\right]dt\right]d\bar{u}$ where $f[\bar{S}_t(\bar{x}), S_t(\psi^{-1}(\bar{x}))]$ is 1 if $\bar{S}_t(\bar{x})$, and $S_t(\psi^{-1}(\bar{x}))$ are in atoms of $\bar{P}$ and P having the same index and 0 otherwise. $\bar{u}$ is the measure on $\bar{X}$ (the point of this remark is that $\bar{d}_\psi$ is small if and only if the names of $\bar{x}$ and $\psi^{-1}(\bar{x})$ are close for most $\bar{x}$).

DEFINITION. Let S_t be a flow. We will call a finite partition P very good if for each x the continuous P-name of x consists of intervals, and the minimum length of these intervals, taken over all x, is > 0. We will denote this minimum by $\ell(P)$.

We will now prove a flow version of the R-K theorem.

LEMMA 1. If S_t is a flow built under a function, then given N and ε, S_t is isomorphic to a flow built under a function, f, with base $\bar{Y}$, where $f \leq N$ and $f = N$ on a set $Y \subset \bar{Y}$ where $m(Y) > (1-\varepsilon)\,m(\bar{Y})$.

We will call $\bigcup_{0 \leq t \leq N} S_tY$ a R-K set of height N with base Y.

Proof. We have already assumed that we can represent the flow S_t as a flow built under a function where Y_1 is the base space, T a transformation on Y_1, and g a function on Y_1.

It is easy to see that we can assume the measure of Y_1 to be as small as we want (by taking a subset of Y_1 of small measure and using that for the base instead of Y_1). Now let A_j be the part of Y_1 where $jN \leq g \leq (j+1)N$. Let $\bar{Y} = \bigcup_{j=0}^{\infty} \bigcup_{i=0}^{i=j} S_{iN}A_j$. We can now represent our flow as a flow built under a function with base $\bar{Y}$ and function $\bar{g}$. Clearly, $\bar{g} \leq N$ and $\bar{g} = N$ except on a set of measure $< m(Y_1)$.

Remark. Let $F_r = \bigcup_{0 \leq t \leq r} S_t Y$. Let P be a finite partition. We can assume the dist (P/F_r) is as close as we want to dist P. (This is obvious if we assume S_1 to be mixing, which we can do in this section with no loss of generality. The remark holds for ergodic flow but requires some argument.)

LEMMA 2. Let S_t and $\overline{S}_t$ be flows built under functions, acting on Lebesgue spaces X and $\overline{X}$. Let P and $\overline{P}$ be partitions of X and $\overline{X}$ such that (a) P generates under S_t*, (b) $\overline{P},\overline{S}_t \sim P,S_t$. Then we can find invariant sets $X_1 \subset X$ and $\overline{X}_1 \subset \overline{X}$ of full measure and a measure- preserving map ψ of $\overline{X}_1$ onto X_1 such that (c) $S_t(\psi(\bar{x}) = \psi(\overline{S}_t(\bar{x}))$ for all $\bar{x}$ in $\overline{X}_1$ (d) $\overline{P}$ and $\psi^{-1}P$ agree a.e. (e) If $\overline{P}$ generates under $\overline{S}_t$, then ψ will be an isomorphism.

Proof. (1) If Q is a partition we will define the smoothing of Q to be the partition $\hat{Q}$ defined as follows: x will be in the ith atom of $\hat{Q}$ if given ε, then for all δ small enough, all t in $(0,\delta)$ except for a set of t of measure $< \varepsilon\delta$, satisfy $S_t(x) \in P^i$. This defines $\hat{Q}$, except on a set of measure 0. We could put this set in $\hat{P}_1$. Because S_t is a flow built under a function we get that the atoms of $\hat{Q}$ are measurable and Q and $\hat{Q}$ agree a.e.

(2) After removing an invariant set of measure 0, we have that any two points have different $\hat{P}$-names.

We see this as follows: Because $\hat{P}$ generates (in the measure algebra sense) under S_t ($\hat{P} = P$ except on a set of measure 0) we have that there is a set $B, m(B) = 0$ such that (a) if x and y are not in B, then they have different $\hat{P}$-names.

We will now change B to $\overline{B}$ such that $\overline{B}$ is invariant, $m(\overline{B}) = 0$ and (a) still holds for $\overline{B}$. We will do this as follows: S_t is a flow built under some function f with some base Y. If $y \in Y$ is such that the measure of the t, $0 \leq t \leq f(y)$ for which $S_t(y) \in B$ has non-zero measure, then we will put $S_t(y)\ -\infty < t < \infty$ in $\overline{B}$.

*In the measure algebra sense (i.e., every measurable set can be approximated arbitrarily well by sets in the algebra generated by a finite number of the $S_t P$).

It is easy to see that $\overline{B}$ is invariant and $m(\overline{B}) = 0$.

If x and y have the same $\hat{P}$-name, $S_t(x)$ and $S_t(y)$ have the same $\hat{P}$-name for all t. Hence, for x (or y) there is a set of t of non-zero measure for which $S_t(x) \in B$. Thus $x \in \overline{B}$.

(3) After removing an invariant set of measure 0 we have that there are a countable number of sets F_i such that (a) F_i is in the completion of the smallest σ-algebra containing all of the $S_t\hat{P}$, (b) any set of measure 0 is contained in a countable union of the $\hat{F}_i$ of arbitrarily small measure.

We see this as follows: Because S_t is a flow built under a function we can find a countable number of sets F_i' satisfying (b) and such that for each i and x the set of t, $S_t(x) \in \hat{F}_i$ is open. Because $\hat{P}$ agrees with P a.e., $\hat{P}$ generates under S_t in the measure algebra sense. Thus we can find sets F_i such that $F_i = F_i'$ a.e. and F_i satisfy (a). It is easy to see that $\hat{F}_i$ contains F_i' except on an invariant set of measure 0.

(4) There are invariant subsets X_1 of X' and $\overline{X}_1$ of $\overline{X}$ of full measure and a map ψ from $\overline{X}_1$ onto X_1 such that the continuous $\hat{\overline{P}}$ name of $\bar{x}$ is the same as the continuous $\hat{P}$ name of $\psi(\bar{x})$.

We will get (4) from (2) and hypothesis (b) by a standard argument as follows:

For each quadruple of rationals (r_1,r_2,r_3,r_4) and atom $\hat{P}^i$ in $\hat{P}$ we get a set $E(r_1,r_2,r_3,r_4,i)$ consisting of those x such that if ℓ is the measure of the set of t, $r_1 < t < r_2$ for which $S_t(x) \in \hat{P}^i$, then $r_3 < \ell < r_4$.

Order the (r_1,r_2,r_3,r_4,i), and denote $E(r_1,r_2,r_3,r_4,i)$ by E_n if (r_1,r_2,r_3,r_4,i) has position n in the above order. Define $\overline{E}_n$ in exactly the same way using $\hat{\overline{P}}$ instead of $\hat{P}$.

(4.1) It is an easy exercise to see that hypothesis (b) implies that if C is any finite collection of integers, then $m\left(\bigcap_{n \in C} E_n\right) = m\left(\bigcap_{n \in C} \overline{E}_n\right)$.

(4.2) It is also easy to see that if x is in E_n and $\bar{x}$ is in $\overline{E}_n$ for exactly the same set of n, then the $\hat{P}$ name of x and the $\hat{\overline{P}}$ name of $\bar{x}$ are the same.

Because of 4.1 and because X and $\overline{X}$ are Lebesgue spaces, we can find $X_1 \subset X'$ and $\overline{X}_1 \subset \overline{X}$ where X_1 and $\overline{X}_1$ have full measure and a measure-preserving map ψ of $\overline{X}_1$ onto X_1 such that for each n and $\bar{x}$ in $\overline{X}_1$ we have that $\bar{x} \in \overline{E}_n$ if and only if $\psi(\bar{x}) \in E_n$ (this follows from a standard argument which we omit).

(4.3) X_1 and $\overline{X}_1$ could be chosen with the additional property that they are invariant.

We see this as follows: If $\psi(\bar{x})$ is defined then we can define $\psi(\overline{S}_t(\bar{x}))$ to be $S_t(\psi(\bar{x}))$. Thus $\overline{S}_t(\bar{x})$ and $\psi(\overline{S}_t(\bar{x}))$ will have the same name. Because of (2) $\psi(\overline{S}_t(\bar{x}))$ is well defined. We have thus enlarged the domain and range of ψ and both are invariant.

(5) Because of (4) and (2) we have that $S_t(\psi(\bar{x})) = \psi(\overline{S}_t(\bar{x}))$ for all $\bar{x}$ in $\overline{X}_1$.

(6) $\hat{\overline{P}}$ and $\psi^{-1}\hat{P}$ agree everywhere. This follows from (4). Hence, $\overline{P}$ and $\psi^{-1}P$ agree a.e.

(7) ψ is measure-preserving.

(a) Let E be a set in the completion of the smallest σ-algebra containing $\hat{P}$. (Call this $\sigma(\hat{P})$). It is then obvious that $\psi^{-1}E$ is in $\sigma(\hat{\overline{P}})$ and $m(E) = m(\psi^{-1}(E))$.

(b) Let E be a set of measure 0. Then given ε there is a collection C of integers such that $m\left(\bigcup_{i \in C} F_i\right) < \varepsilon$ and $E \subset \bigcup_{i \in C} \hat{F}_i$. Hence

$$\psi^{-1}E \in \bigcup_{i \in C} \psi^{-1}\hat{F}_i = \bigcup_{i \in C} \left(\widehat{\psi^{-1}F_1}\right)$$

(because of 4). Thus $\psi^{-1}E$ is measurable and has measure 0.

(c) Because $\hat{P}$ generates under S_t in the measure algebra sense, any measurable set E is equal a.e. to a set satisfying (a). This and (b) give (7).

(8) If $\overline{P}$ generated under $\overline{S}_t$, then (2) and (3) would have held for $\overline{P}$ as well as P. We get that ψ is 1 - 1 from (2) and that ψ^{-1} is measurable, from (3) and (7).

Part II

DEFINITION. Let S_t be a flow on X and P a partition. We say that <u>x is ε good relative to δ and P (or ε good rel δ,P)</u> if all but $\varepsilon\delta$ of the t between 0 and δ are such that $S_t(x)$ and $x = S_0(x)$ lie in the same atom of P.

We say that <u>P is ε good rel δ</u> if all but ε of the x in X are ε good rel δ,P.

We say that the continuous P-name of x of length n is ε good rel δ if there exists a subset F of (0,n) of measure $> (1-\varepsilon)n$ such that if $t \in F$ then $S_t(x)$ is ε good rel δ,P.

LEMMA 3. Let S_t be a flow built under a function, and P a partition. Then given ε there exists a $\bar{\delta}$ such that P is ε good rel δ for all $0 \leq \delta \leq \bar{\delta}$.

Proof. This is a standard theorem in real variables (since S_t is a flow built under a function).

DEFINITION. Let S_t be a flow and P a partition. By the δ filled P-name of x, of length r we will mean the partition of 0,r whose i^{th} atom consists of those t such that $k\delta \leq t \leq (k+1)\delta$, $0 \leq t \leq r$, and $S_{k\delta}(x)$ lies in the ith atom of P.

LEMMA 4. Let S_t and $\bar{S}_t$ be flows built under functions acting on X and $\bar{X}$. Let P and $\bar{P}$ be partitions of X and $\bar{X}$ such that (a) P and $\bar{P}$ are both ε good rel $\frac{1}{M}$. If (b) $\bar{d}\left(\left\{S^i_{\frac{1}{M}}P\right\}_0^{nM}, \left\{\bar{S}^i_{\frac{1}{M}}\bar{P}\right\}_0^{nM}\right) < \alpha$, then

(c) $\bar{d}\left(\{S_tP\}_0^n, \{\bar{S}_t\bar{P}\}_0^n\right) < \alpha + 5\varepsilon$.

Proof. (1) Because of (b) there is a φ, mapping X onto $\bar{X}$ such that if we let $f(x) = \frac{1}{n} \cdot \{$number of i between 0 and nM such that $S^i_{\frac{1}{M}}(x)$ and $\bar{S}^i_{\frac{1}{M}}\varphi(x)$ lie in atoms of P and $\bar{P}$ having different indices$\}$, then $\int f(x)\,dx < \alpha$. Note that f(x) measures the disagreement between the P-name of x (under $S_{\frac{1}{M}}$ of length nM) and the $\bar{P}$-name of $\varphi(x)$ (under $\bar{S}_{\frac{1}{M}}$ of length nM). Thus the P-name of x and the $\bar{P}$-name of $\varphi(x)$ agree to within ε on the average.

(2) Because of (a) we have that the P-name of x under $S_{\frac{1}{M}}$ of length nM and the $\frac{1}{M}$ filled P-name of x of length n agree to within 2ε on the average. (The same holds for $\bar{P}$.)

(3) Because of (1) we have that the $\frac{1}{M}$ filled P-name of x and the $\frac{1}{M}$ filled $\overline{P}$-name of $\varphi(x)$ agree to within ε on the average.

(4) Our lemma follows easily from (2) and (3).

The following lemma is an extension to flows of the fact that any partition of a Bernoulli shift is F.D.

LEMMA 5. Let S_t and $\overline{S}_t$ be flows built under functions acting on X and $\overline{X}$. Let $\overline{S}_1$ be a Bernoulli shift of finite entropy. Let $\overline{P}$ be a finite partition of $\overline{X}$, such that $\overline{P}$ is ε good rel $\frac{1}{M}$, then there exists δ and u such that if $\tilde{P}$ satisfies: (a) $|\tilde{P}| = |\overline{P}|$, (b) $\left|\operatorname{dist}\left(\bigvee_0^u S^i_{\frac{1}{M}} \tilde{P}\right) - \operatorname{dist}\left(\bigvee_0^u \overline{S}_{\frac{1}{M}} \overline{P}\right)\right| < \delta$, (c) $\left|E\left(\tilde{P}, S_{\frac{1}{M}}\right) - E\left(\overline{P}, \overline{S}_{\frac{1}{M}}\right)\right| < \delta$, and (d) $\tilde{P}$ is 2ε good rel $\frac{1}{M}$, then (e) $\overline{d}((S_t\tilde{P}), (\overline{S}_t\overline{P})) < 10\varepsilon$.

Proof. Since roots of Bernoulli shifts are Bernoulli (pp. 26, 27), we have that $\overline{S}_{\frac{1}{M}}$ is a Bernoulli shift. Since any partition of a B-shift is F.D. (p. 24), we get that $\overline{P}$ is F.D. relative to $\overline{S}_{\frac{1}{M}}$. Thus we can choose u and δ so that (a), (b), and (c) imply (f) $\overline{d}\left[\left(\tilde{P}, S_{\frac{1}{M}}\right), \left(\overline{P}, \overline{S}_{\frac{1}{M}}\right)\right] < \varepsilon$. Because of the previous lemma (f) and (d) imply (e).

LEMMA 6. Let S_t be a flow built under a function and P a partition. Given ε, there exists an M' with the following property: If $M > M'$, and if $\bar{\varepsilon} > 0$, then for all n large enough we will have that for all x, except for a set of measure $< \bar{\varepsilon}$, the $\frac{1}{M}$ filled name (and the continuous name) of x of length n will be ε good rel $\frac{1}{M}$.

Proof. Choose M' so small that for all $M > M'$, P is $\bar{\varepsilon}$ is good rel $\frac{1}{M}$. Fixing M, we can choose n so large that there is a set F, $m(F) > 1 - \bar{\varepsilon}$, and for $x \in F$ (a) the $\frac{1}{M}$ filled P-name of x of length n differs from the continuous P-name of x, of length n in a set of measure $< 2\varepsilon$, and (b) the continuous P-name (of length n) of x is $\frac{1}{2}\varepsilon$ good rel $\frac{1}{M}$.

If $(\bar{\varepsilon})^{\frac{1}{2}}$ is small compared to ε, then (a) and (b) will imply that the $\frac{1}{M}$ filled name of x will be ε good rel $\frac{1}{M}$.

DEFINITION. Let S_t be a flow and P a finite partition. We will say that $\varepsilon, \frac{1}{M}$ works for P if P is ε good rel $\frac{1}{M}$ and if given $\bar{\varepsilon}$ then for all n large enough we will have that except for a collection of x of measure $< \bar{\varepsilon}$, the $\frac{1}{M}$ filled name (and the continuous name) of x of length n will be ε good rel $\frac{1}{M}$.

The following is the main lemma.

LEMMA 7. Let $\bar{S}_t$ be a flow such that $\bar{S}_1$ is a Bernoulli shift (of finite entropy). Let $\bar{P}$ be a finite partition that generates under $\bar{S}_1$. Let S_t be a flow such that S_1 is mixing and $E(S_1) = E(\bar{S}_1)$. Let P be a finite partition satisfying [1] $\bar{d}((P,S_t),(\bar{P},\bar{S}_t)) < \left(\frac{\varepsilon}{100}\right)^2$.

Then given $\tilde{\varepsilon}$ there is a $\tilde{P}$ such that $[\tilde{0}]$ $|\tilde{P} - P| < \varepsilon$,

$[\tilde{1}]$ $\bar{d}((\tilde{P},S_t),(\bar{P},\bar{S}_t)) < \tilde{\varepsilon}$.

<u>Proof.</u> (1) It will be convenient to assume $\tilde{\varepsilon} < \frac{1}{100}\varepsilon$. It will also be convenient (but not necessary) to assume that P is finite and very good (we can do this because we can modify P on an arbitrarily small set to achieve this without disturbing [1]).

Apply Lemma 6 to pick M so that $\frac{1}{10}\tilde{\varepsilon}, \frac{1}{M}$ works for P, and $\bar{P}$. Apply Lemma 5 to get $\delta > 0$ and u such that if $\tilde{P}$ satisfies (a) $|\bar{P}| = |\tilde{P}|$,

(b) $\left|\operatorname{dist}\left(\bigvee_0^u S_{\frac{1}{M}}^i \tilde{P}\right) - \operatorname{dist}\left(\bigvee_0^u \bar{S}_{\frac{1}{M}} \bar{P}\right)\right| < \delta$, (c) $\left|E\left(\tilde{P},S_{\frac{1}{M}}\right) - E\left(\bar{P},\bar{S}_{\frac{1}{M}}\right)\right| < \delta$,

(d) $\tilde{P}$ is $\frac{1}{5}\varepsilon$ good rel $\frac{1}{M}$, (e) $|\tilde{P} - P| < \varepsilon$, then $\tilde{P}$ will satisfy the conclusions of our lemma.

(2) Choose a very good finite partition Q such that $\left|E(Q,S_{\frac{1}{M}}) - E(S_{\frac{1}{M}})\right| < \frac{1}{100}\delta$.

Choose δ so that if $|Q_1| = |Q|$ and $|Q_1 - Q| < \hat{\delta}$, then there is a partition R,$E(R) < \frac{1}{100}\delta$ and $Q_1 \vee R \supset Q$. It will be convenient to choose $\gamma < \frac{1}{100}\delta$, $\gamma < \frac{1}{100}\hat{\delta}$, $\gamma < \frac{1}{100}\varepsilon$, $\gamma < \frac{1}{100}\tilde{\varepsilon}$. (We will show that $\tilde{P}$ has good entropy by showing that it catches Q to within γ.)

(3) Choose L to be a multiple of M such that $\frac{1}{L} < \frac{1}{100}\gamma\ell(P)\frac{1}{L} < \frac{1}{100}\gamma\ell(Q)$.

(4) For any n we can find partition $\hat{P}_i$ such that

$$\text{(a) } \frac{1}{nM}\sum_{i=-nM}^{i=-1}\left|\hat{P}_i - S^i_{\frac{1}{M}}(P)\right| < \left(\frac{\varepsilon}{100}\right)^2 \quad \text{(b) dist } \bigvee_{-nM}^{-1}\hat{P}_i = \text{dist } \bigvee_{-nM}^{-1}\bar{S}^i_{\frac{1}{M}}\bar{P}.$$

This follows immediately from hypothesis [2].

Let us establish some terminology. We will call an atom in $\bigvee_{-nL}^{-1} S^i_{\frac{1}{M}} P$ (or $\bigvee_{-nL}^{-1} S^i_{\frac{1}{L}}(P \vee Q)$) an n-P-atom (or n-$P \vee Q$-atom). $\bigvee_{-nL}^{-1} S^i_{\frac{1}{L}} P$ partitions X according to the P-name of points under $S_{\frac{1}{L}}$ of length nL. We will refer to these names as n-P-names. We will define the n-$\hat{P}$-name of x to be the sequence whose i^{th} term is the index of the atom of $\hat{P}_i$ containing x. Let B_j be an atom of $\bigvee_{-nM}^{-1}\hat{P}_i$. Because of (b) there is a point, $\bar{x}(B_j)$ (in $\bar{X}$, not unique), such that the n-$\hat{P}$-name of any x in A_j is the same as the $\bar{P}$-name under $\bar{S}_{\frac{1}{M}}$ of length nM of $\bar{x}(B_j)$.

(5) We will now show that we can (for all n large enough) find a collection of the n-$P \vee Q$-atoms and a collection of the n-$\hat{P}$-atoms with certain good properties. We will call the atoms in the first collection good n-$P \vee Q$-atoms and those in the second good n-$\hat{P}$-atoms.

We will now list the properties of good atoms:

(a) The good n-$\hat{P}$-atoms cover all but $(1 - \frac{1}{100}\gamma)$ of X;

The good n-$P \vee Q$-atoms cover all but $(1 - \frac{1}{100}\gamma)$ of X;

(b) The good n-$\hat{P}$-atoms are not much larger than the good n-$P \vee Q$-atoms. The exact statement of the above is the following: <u>The measure of any good</u>

n-P ∨ Q-atom is greater than $\frac{1}{2}^{\frac{1}{10}\delta n}$ times the measure of any good n-$\hat{P}$-atom.

We get the above as follows: If n is large enough, then all but $\frac{1}{100}\gamma$ of the n-P ∨ Q-atoms have measure $> \frac{1}{2}^{[E(S_{\frac{1}{L}}) + \frac{\delta}{100L}]nL} = \frac{1}{2}^{[E(S_1) + \frac{\delta}{100}]n}$.

(This follows from the S.M.B. theorem.) Similarly, all but $\frac{1}{100}\gamma$ of the n-$\hat{P}$-atoms have measure $< \frac{1}{2}^{[E(\overline{P},S_{\frac{1}{M}}) - \frac{\delta}{100L}]nM} = \frac{1}{2}^{[E(S_1) + \frac{\delta}{100}]n}$

$= \frac{1}{2}^{[E(S_1) + \frac{\delta}{100}]n} \cdot \frac{1}{2}^{-\frac{3}{100}\delta n}$.

(c) Each n-$\hat{P}$-atom has a name with nM terms. We will require that if the n-$\hat{P}$-atom is good, then the distribution of u names (subsequences of u consecutive terms) in its name is within $\frac{1}{100}\delta$ of the distribution of $\overline{P}$-names of length u under $\overline{S}_{\frac{1}{M}}$ in $\overline{X}$.

We can require this because the names of the n-$\hat{P}$-atoms are distributed in the same way as the $\overline{P}$-names under $\overline{S}_{\frac{1}{M}}$, of length nM of points in $\overline{X}$. Because of the ergodic theorem we can assume that if n is large enough, the distribution of $\overline{P}$-names under $\overline{S}_{\frac{1}{M}}$ of length u in the $\overline{P}$-name x will have distribution arbitrarily close to the distribution of $\overline{P}$-names under $\overline{S}_{\frac{1}{M}}$ of length u in $\overline{X}$ for all but an arbitrarily small set of x.

(d) We will require if B_j is a good n-$\hat{P}$-atom then $x(B_j)$ has the property that the $\frac{1}{M}$ filled $\overline{P}$-name of $x(B_j)$ is $\frac{1}{9}\tilde{\varepsilon}$ good rel $\frac{1}{M}$ (note that the above condition does not depend on the particular choice of $x(B_j)$). We can require this because $\frac{1}{10}\tilde{\varepsilon}$, $\frac{1}{M}$ works for $\overline{P}$.

(e) If α_j is a good n-P ∨ Q-atom and if $x \in \alpha_j$, then the $\frac{1}{L}$ filled P-name of x and length n and the $\frac{1}{M}$ filled P-name of x of length n differ for a set of t of measure $< \frac{1}{100}\varepsilon$.

We get this because $\tilde{\varepsilon} < \frac{1}{100}\varepsilon$, and because if $L > M$ and if n is large enough, then for an arbitrarily large set of x we have that the $\frac{1}{L}$ filled P-name of length n of x and the $\frac{1}{M}$ filled P-name of x of length n each differ from the continuous P-name of x of length n for a set of t of measure $< \frac{1}{100}\varepsilon$. We get this because P is $\frac{1}{10}\tilde{\varepsilon}$ good rel $\frac{1}{M}$ and P is $\frac{1}{10}\tilde{\varepsilon}$ good rel $\frac{1}{L}$.

(6) We will now use 5b to assign to each good n-$P \vee Q$-atom, α_i, a good n-$\hat{P}$-atom, B_j. Each good n-$\hat{P}$-atom will be used at most $2^{\frac{1}{10}\delta n}$ times. Furthermore, all but $\frac{1}{100}\varepsilon$ of the good n-$P \vee Q$-atoms, α_i will be assigned a good n-$\hat{P}$-atom B_j such that the n-$\hat{P}$-name of B_j agrees with the P-name under $S_{\frac{1}{M}}$ of length nM of any point in α_i in $> (1 - \frac{1}{100}\varepsilon)$ places. (This is equivalent to saying that if $x \in \alpha_i$, then the $\frac{1}{M}$ filled P-name of x of length n agrees with the $\frac{1}{M}$ filled $\bar{P}$-name of $\bar{x}(B_j)$ of length n for all $t(0 \leq t \leq n)$ except a set of measure $< \frac{1}{100}\varepsilon n$.)

Because $\frac{1}{nM} \sum_{-nM}^{-1} \left| \hat{P}_i - S^i_{\frac{1}{M}} P \right| < (\frac{\varepsilon}{100})^2$, we have that there is a set $E, m(E) > 1 - \frac{\varepsilon}{100}$ such that if $x \in E$, then the n-$\hat{P}$-name of x and the n-P-name of x (under $S_{\frac{1}{M}}$) agree in more than $(1 - \frac{\varepsilon}{100})nM$ places. Let E' be the part of E covered by good n-$P \vee Q$-atoms.

We will call an n-$P \vee Q$-atom very good if more than half of it lies in E'. Call the part of E' thus covered E''.

If we divide each good n-$\hat{P}$-atom into $2^{\frac{1}{10}\delta n}$ equal parts, then (by 5b) each of these parts will be smaller than half of a good n-$P \vee Q$-atom. We can therefore apply the marriage lemma on E'' and assign to each very good n-$P \vee Q$-atom a part of a n-$\hat{P}$-atom that intersects it in E''. We can now assign to the rest of the good n-$P \vee Q$-atoms a part of a good n-$\hat{P}$-atom not already used.

(7) We now form a continuous R-K set of height n filling all but $\frac{1}{100}\gamma$ of the space. Let F_L be the part of height $\leq \frac{1}{L}$ (that is, if Y is its base then

$F_L = \bigcup_{0 \leq t \leq \frac{1}{L}} S_t Y$ and $\bigcup_{0 \leq t \leq n} S_t Y = \bigcup_{i=0}^{nL} S^i_{\frac{1}{L}} F_L$). Because S_1 is mixing, we

can assume that the very good n-$P \vee Q$-atoms cover all but $\frac{2}{100}\varepsilon$ of F_L and the good n-$P \vee Q$-atoms cover all but $\frac{2}{100}\gamma$ of F_L. The parts of the n-$P \vee Q$-atoms that lie in F_L we will call n-$P \vee Q$-F-atoms. From now on A_i will always denote a good, or very good, n-$P \vee Q$-F-atom.

(8) A set in F_L we will call Y-measurable if each set of the form $\bigcup_{0 \leq t \leq \frac{1}{L}} S_t(y)$, y in Y, is either in the set or its compliment.

We will now modify each A_i to obtain a Y-measurable set A_i'. We will do this as follows: Order the A_i so that the very good ones come first. Suppose we have defined A_i' for $1 \leq i \leq r$ so that the A_i' are disjoint and $\bigcup_1^r A_i' \supset \bigcup_1^r A_i$. We then look at the part, $\tilde{A}_{r+1}$ of A_{r+1} that lies in $F_L - \bigcup_1^r A_i'$, and let A'_{r+1} be the smallest Y-measurable set in $F_L - \hat{\bigcup}_1 A_i'$ containing $\tilde{A}_{r+1}$.

(9) We will now define $\tilde{P}$ and auxiliary sets Q'' and P''.[1]

(a) If $y \in Y \cap A_j'$, then the continuous Q''-name of length n of y will be the $\frac{1}{L}$ filled Q-name (of length n) of any $x \in A_j$. (Thus the Q''-name under $S_{\frac{1}{L}}$ of length nL is the same for all x in A_j' and equal to the Q-name under $S_{\frac{1}{L}}$ of any point in A_j. Furthermore, the $S^i_{\frac{1}{L}} A_j'$, $0 \leq i \leq nL$ will refine Q'' on $\bigcup_{i=0}^{nL} S^i_{\frac{1}{L}} (\bigcup A_j')$.)

(b) If $y \in A_j' \cap Y$, then the continuous P''-name of y of length n will be the same as the $\frac{1}{L}$ filled P-name of any x in A_j.

(c) If $y \in A_j' \cap Y$ and if B_i is the n-$\hat{P}$-atom assigned to A_j, then the continuous $\tilde{P}$-name of y of length n will be the same as the $\frac{1}{M}$ filled $\overline{P}$-name of $\bar{x}(B_i)$ (under $\overline{S}_t$ of length n).

(10) $|P'' - P| < \frac{2}{100}\gamma$. We see this as follows: All points in A'_j have the same P''-name under $S_{\frac{1}{L}}$ of length nL and this is the n-P-name of the points in A_j. Because of the construction of A'_j we have that if $x \in A'_j$, then for some t, $|t| < \frac{1}{L}$, $S_t x \in A_j$. Thus, the n-P''-name of x is the same as the n-P-name of $S_t(x)$ for some t, $|t| \leq \frac{1}{L}$. But it is easy to see from the choice of L that the n-P-name of x and $S_t(x)$ differ in $< \frac{1}{100}\gamma\, nL$ places. Thus, for all $x \in \bigcup A'_j$ we have that the P''-name and the P-name (under $S_{\frac{1}{L}}$ of length nL) differ in $< \frac{1}{100}\gamma\, nL$ places.

The same argument gives $|Q'' - Q| < \frac{2}{100}\gamma$.

(11) We will now show that $|\tilde{P} - P| < \varepsilon$. Because of (10), it will be enough to show that $|\tilde{P} - P''| < \frac{1}{2}\varepsilon$. We do this as follows: Fix a very good A_j and let B_i be the n-$\hat{P}$-atoms assigned to A_j. Let y be some point in $A'_j \cap Y$.

(a) By (9c) the continuous $\tilde{P}$-name (length n) of y equals the $\frac{1}{M}$ filled $\overline{P}$-name of $\bar{x}(B_i)$.

(b) By (6) the $\frac{1}{M}$ filled $\overline{P}$-name of $\bar{x}(B_j)$ and the $\frac{1}{M}$ filled name of any $x \in A_j$ differ for a set of t of measure $< \frac{1}{100}\varepsilon$.

(c) By (5e) the $\frac{1}{M}$ filled P-name of $x \in A_j$ and the $\frac{1}{L}$ filled P-name of x (both of length n) disagree for a set of t of measure $< \frac{1}{100}\varepsilon$.

(d) By (9b) the $\frac{1}{L}$ filled P-name of $x \in A_j$ equals the continuous P''-name of y.

(e) Because of (a) – (d) and because $\bigcup_{i=0}^{nL} S^i_{\frac{1}{L}}\left(\bigcup_{\text{very good } A_j} A_j\right)$ fills all but $\frac{3}{100}\varepsilon$ of X, we get (11).

(12) We will now check that $\left| E(\tilde{P}, S_{\frac{1}{M}}) - E(\overline{P}, \overline{S}_{\frac{1}{M}}) \right| < \delta$. Because of our choice of Q, Q'', γ, and (10), it will be enough to show that we can find a partition R, $E(R) < \frac{1}{2}\delta$ such that $\bigvee_{-nM}^{nM} S^i_{\frac{1}{M}}(\tilde{P} \vee R) \overset{\gamma}{\supset} Q''$.

Because $< 2^{\frac{1}{10}\delta n}$ good n-$\hat{P}$-atoms were assigned to each A'_j, we can find a partition R_1, $|R_1| < 2^{\frac{1}{10}\delta n}$, such that R_1, together with the partition of

$(\cup A_j)$ according to the $\tilde{P}$-names under $S_{\frac{1}{M}}$ of length nM of the points in $(\cup A_j)$, will determine the A_j.

If $0 \le i \le \frac{L}{M}$ and if x is in $S^i_{\frac{1}{L}}(\cup A_j)$, then we can tell which $S^i_{\frac{1}{L}} A_j$, x lies in by knowing which atom of $S^i_{\frac{1}{L}} R_1$, x lies in and the $\tilde{P}$-names of x under $S_{\frac{1}{M}}$ of length nM.

We can therefore let R be the partition whose atoms consist of the $S^i_{\frac{1}{L}} R^j_1$. $0 \le i \le \frac{L}{M}$, R^j_1 being the atom of R_1. Since R partitions a set of measure $< \frac{1}{n}$ into $< 2^{\frac{1}{5}\delta n}$ parts we get that (for all n large enough) $E(R) < \frac{1}{2}\delta$.

(13) We will now check that $\left| \operatorname{dist}\left(\bigvee_0^u S^i_{\frac{1}{M}} P\right) - \operatorname{dist}\left(\bigvee_0^u \bar{S}^i_{\frac{1}{M}} \bar{P}\right) \right| < \delta$.

We get this because by 5c n-$\hat{P}$-names of good atoms had this property, and hence by (9c) the $\tilde{P}$-names of length n of the y in $Y \cap (\cup A_j)$ have this property. Since $\bigcup_{0\le t\le n} S_t(Y \cap (\cup A_j))$ fills all but $\frac{1}{2}\delta$ of X, we get (13).

(14) We must show that $\tilde{P}$ is $\frac{1}{5}\tilde{\varepsilon}$ good rel $\frac{1}{M}$. Because of (9c), the continuous $\tilde{P}$-name of y, $y \in Y \cap (\cup A_j)$, is the same as the $\frac{1}{M}$ filled $\bar{P}$-name of $\bar{x}(B_i)$. By (5d) we get that this name is $\frac{1}{9}\tilde{\varepsilon}$ good rel $\frac{1}{M}$. Since $\bigcup_{0\le t\le n} S_t((\cup A_j)\cap Y)$ fills all but $\frac{1}{10}\tilde{\varepsilon}$ of X, we get (14).

PROPOSITION 8. Let $\bar{S}_t$ be a Bernoulli flow such that $E(\bar{S}_1) < \infty$. Let $\bar{P}$ be a finite partition that generates under $\bar{S}_1$. Let S_t be a flow that S_1 is mixing and $E(S_1) = E(\bar{S}_1)$.

Let P be a finite partition satisfying [1] $\bar{d}((P,S_t),(\bar{P},\bar{S}_1)) < (\frac{\varepsilon}{100})^2$.

Then there is a partition $\tilde{P}$ such that [2] $|\tilde{P} - P| < \varepsilon$, [3] $\tilde{P},S_t \sim \bar{P},\bar{S}_t$.

Proof. This follows from repeated applications of the previous lemma.

COROLLARY 9. Let $\bar{S}_t$ be a Bernoulli flow such that $E(\bar{S}_1) < \infty$. Let $\bar{P}$

be a finite partition that generates under $\bar{S}_1$. Let S_t be a flow such that S_1 is mixing and $E(S_1) = E(\bar{S}_1)$.

Then there exists a partition $\tilde{P}$ such that $\tilde{P},S_t \sim \bar{P},\bar{S}_t$.

Proof. This is a consequence of the previous proposition. If we let $\varepsilon = 100$, then [2] becomes vacuous, and we can always find a good partition P. (If we deduce Corollary 9 from the main lemma, it goes like this: We first prove the main lemma for the case $\varepsilon = 100$. In this case P plays no role, and we are simply required to produce $\tilde{P}$ without having the additional problem of keeping it close to a previously given P. We then apply the full lemma an infinite number of times to obtain our corollary.)

For part III of the isomorphism theorem it will be convenient to state the main result of part II in the following form:

PROPOSITION 10. Let $\bar{S}_t$ be a flow such that $\bar{S}_1$ is a Bernoulli shift of finite entropy. Let $\bar{P}$ be a finite partition that generates under $\bar{S}_1$. Let M be an integer such that $\bar{P}$ is $\frac{1}{10}\varepsilon$ good rel $\frac{1}{M}$. Then given ε there is a $\delta > 0$ and an integer u such that if P is a finite partition under S_t (where S_1 is mixing) satisfying (a) P is $\frac{1}{5}\varepsilon$ good rel $\frac{1}{M}$, (b) $|P| = |\bar{P}|$,

(c) $\bar{d}\left(\left\{S^i_{\frac{1}{M}}P\right\}_{i=1}^{u}, \left\{\bar{S}^i_{\frac{1}{M}}\bar{P}\right\}_{i=1}^{u}\right) < \delta$, (d) $\left|E(P,S_{\frac{1}{M}}) - E(\bar{P},\bar{S}_{\frac{1}{M}})\right| < \delta$,

then there is a $\tilde{P}$ such that (e) $|\tilde{P} - P| < \varepsilon$, (f) $\tilde{P},S_t \sim \bar{P},\bar{S}_t$.

Proof. Using Lemma 6 we can choose δ, u, and M so that (a), (b), (c), and (d) imply $\bar{d}((P,S_t),(\bar{P},\bar{S}_t)) < (\frac{\varepsilon}{100})^2$. We now apply Proposition 8 to get $\tilde{P}$.

Part III

DEFINITION. We will define a continuous gadget (or a gadget for S_t) to be an R-K set of height n and base Y together with a partition P of $\bigcup_{0\leq t\leq n} S_tY$. We will say that the gadget of height $\bar{n}$ and base $\bar{Y}$ with partition $\bar{P}$ is isomorphic to the first gadget if $\bar{n} = n$ and the distribution of continuous

P-names of length n in Y is the same as the distribution of continuous $\overline{P}$-names of length n in $\overline{Y}$.

LEMMA 11. Let S_t be a Bernoulli flow such that $E(S_1) < \infty$. Let P and Q be finite partitions such that P generates under S_1. Assume that there is a σ-algebra X_Q, generated by Q under S_1 and invariant under S_t, and that S_t acting on X_Q is a flow (i.e., is a flow built under a function). We also assume that (a) $E(P,S_1) = E(Q,S_1)$.

Then given ε there is a Q_1 such that (1) $Q_1,S_t \sim Q,S_t$, (2) $\bigvee_{\infty}^{\infty} S_1^i Q_1 \overset{\varepsilon}{\supset} P$, (3) $|Q_1 - Q| < \varepsilon$.

Proof. Note that because of Corollary 3, p. 20, factors of Bernoulli shifts are Bernoulli, and hence S_t acting on X_Q is a Bernoulli flow. The first part of the proof will be devoted to proving the following: (A) There is a partition $\overline{P}$ such that (i) $\overline{P} \subset \bigvee_{-\infty}^{\infty} S_t Q$, (ii) $\overline{P},S_t \sim P,S_t$, and (iii) there is a K and an operator L (see the definition on page 27) such that $\left| L\left(\bigvee_{-K}^{K} S_1^i P\right) - Q\right| < \varepsilon$ and $\left| L\left(\bigvee_{-K}^{K} S_1^i \overline{P}\right) - Q\right| < \varepsilon$.

We get (A) as follows: (1) Because P generates under S_1, there is a K and an operator L such that (1.1) $\left| L\left(\bigvee_{-K}^{K} S_1^i P\right) - Q\right| < \frac{1}{100}\varepsilon$. Pick M so that P is $\frac{\varepsilon}{100K}$ good rel $\frac{1}{M}$. Apply Proposition 10 to S_t acting on X_Q to get $\delta > 0$, an n such that if P' satisfies (a) P' is $\frac{2\varepsilon}{100K}$ good rel $\frac{1}{M}$, (b) $|P'| = |P|$, (c) $\bar{d}(\{S_{\frac{1}{M}}^i P\}_1^u, \{S_{\frac{1}{M}}^i P'\}_1^u) < \delta$, (d) $\left|E(P',S_{\frac{1}{M}}) - E(P,S_{\frac{1}{M}})\right| < \delta$, then there exists a $\overline{P}$ such that (e) $\overline{P} \subset \bigvee_{-\infty}^{\infty} S_t Q$, (f) $|\overline{P} - P'| < \frac{\varepsilon}{10K}$, (g) $\overline{P},S_t \sim P,S_t$.

Choose $\bar{\delta}$ such that $\bar{\delta} < \delta$ and $\bar{\delta} < \varepsilon$, and if $|Q^{\infty} - Q| < \bar{\delta}$, then $\left|E(Q^{\infty},S_{\frac{1}{M}}) - E(Q,S_{\frac{1}{M}})\right| < \delta$.

Choose K_1 so that $K_1 > K$, $K_1 > u$ and (1.2) $\bigvee_{-K_1}^{K_1} S_1^i P \overset{\frac{1}{10}\bar{\delta}}{\supset} Q$.

Choose N so that $K_1 < \frac{\bar{\delta}}{100}N$.

(2) Apply Lemma 1 to S_t acting on $\bar{X}_Q$ to find a continuous R-K set of height N that fills up X_Q to within $\frac{\bar{\delta}}{100}$. Let Y be the base.

Let $\hat{Q}$ be a good finite partition in X_Q such that $|\hat{Q} - Q| < \frac{1}{10}\bar{\delta}$ and such that the points in Y have only a finite number of different continuous $\hat{Q}$-names of length N.

Let G_1 be the gadget formed by $\bigcup_0^N S_t Y$ and $P \vee \hat{Q}$. Choose P' in X_Q such that if G_2 is formed by $\bigcup_0^N S_t Y$, and $P' \vee \hat{Q}$, then $G_1 \sim G_2$. This defines P' on $\bigcup_0^N S_t Y$. It does not matter how we define P' on the rest of the space.

(3) Because $|Q - \hat{Q}| < \frac{1}{10}\bar{\delta}$, (1.1) and (1.2) give $\left| L\left(\bigvee_{-K}^{K} S_1^i P \right) - \hat{Q} \right| < \frac{2}{10}\varepsilon$ and $\bigvee_{-K_1}^{K_1} S_1^i P \overset{\frac{2}{10}\bar{\delta}}{\supset} \hat{Q}$. Since $G_1 \sim G_2$* and fills most of X and $K < K_1 < \frac{\bar{\delta}}{100}N$, we get $\left| L\left(\bigvee_{-K}^{K} S_1^i P' \right) - \hat{Q} \right| < \frac{3}{10}\varepsilon$ and $\bigvee_{-K_1}^{K_1} S_1^i P' \overset{\frac{3}{10}\bar{\delta}}{\supset} \hat{Q}$. Again, because $|Q - \hat{Q}| < \frac{1}{10}\bar{\delta}$, we get (3.1) $\left| L\left(\bigvee_{-K}^{K} S_1^i P' \right) - Q \right| < \frac{4}{10}\varepsilon$, (3.2) $\bigvee_{-K_1}^{K_1} S_1^i P' \overset{\frac{4}{10}\bar{\delta}}{\supset} Q$.

Because $G_1 \sim G_2$ and fill most of the space, we have that $\left| \operatorname{dist}\left(\bigvee_0^u S_{\frac{1}{M}}^i P' \right) - \operatorname{dist}\left(\bigvee_0^u S_{\frac{1}{M}}^i P \right) \right| \leq \bar{\delta} \leq \delta$. Because of (3.2) and the choice of $\bar{\delta}$, we $|E(P', S_{\frac{1}{M}}) - E(P, S_{\frac{1}{M}})| < \delta$. Since $G_1 \sim G_2$ and fill most of the space, and since P is $\frac{\varepsilon}{100K}$ good rel $\frac{1}{M}$, we have P' is $\frac{\varepsilon}{100K}$ good rel $\frac{1}{M}$. Because of the above inequalities and because of our choice of u and δ, we get $\bar{P}$ such that (a) $\bar{P} \subset \bigvee_{-\infty}^{\infty} S_1^i Q$, (b) $\bar{P}, S_t \sim P, S_t$, (c) $|\bar{P} - P'| < \frac{\varepsilon}{10K}$.

*This can be seen in exactly the same way as in the discrete time case.

Because of (c) and (3.1), we get (d) $\left|L\left(\bigvee_{-K}^{K} S_1^i \overline{P}\right) - Q\right| < \varepsilon$. (a), (b), and (d) give (A) (since ε is arbitrary).

We will now use (A) to prove our lemma.*

(4) Because $\overline{P} \subset \bigvee_{-\infty}^{\infty} S_t Q = \bigvee_{-\infty}^{\infty} S_1^i Q$ (see the hypothesis of our lemma), we can find K_2 such that (4.1) $\bigvee_{-K_2}^{K_2} S_1^i Q \overset{\varepsilon/10}{\supset} \overline{P}$.

As before, we will pick M so that Q is $\frac{\varepsilon}{100K_2}$ good rel $\frac{1}{M}$ and apply Proposition 10 to find $\delta > 0$ and u such that if Q' satisfies (a) Q' is $\frac{2\varepsilon}{100K}$ good rel $\frac{1}{M}$, (b) $|Q| = |Q'|$, (c) $\bar{d}(\{S_{\frac{1}{M}}^i Q\}_1^u, \{S_{\frac{1}{M}}^i Q'\}) < \delta$, (d) $\left|E(P', S_{\frac{1}{M}}) - E(P, S_{\frac{1}{M}})\right| < \delta$, then there exists $\overline{Q}$ such that $|\overline{Q} - Q| < \frac{\varepsilon}{10K_2}$ and $\overline{Q}, S_t \sim Q, S_t$.

Choose $\bar{\delta}$ ($\bar{\delta} < \delta$, $\bar{\delta} < \varepsilon$) so that if $|P^{\infty} - P| < \bar{\delta}$, then $\left|E(P^{\infty}, S_{\frac{1}{M}}) - E(P, S_{\frac{1}{M}})\right| - E(P, S_{\frac{1}{M}})\Big| < \delta$.

Choose K_3 so that $K_3 > K_2$, $K_3 > u$ and (4.2) $\bigvee_{-K_3}^{K_3} S_1^i Q \overset{\frac{1}{10}\bar{\delta}}{\supset} \overline{P}$.

Choose N so that $K_3 < \frac{\bar{\delta}}{100} N$ and $\frac{1}{N} \ell(\delta', Q) < \frac{1}{10}\delta'$.

(5) We now form a continuous R-K set of height N and base Y_1 that fills X to within $\bar{\delta}/100$.

Let G_1'' be the gadget formed by partitioning the above R-K set by P. Because $P, S_t \sim \overline{P}, S_t$, we can find a gadget G_2'' in $\bigvee_{-\infty}^{\infty} S_t \overline{P}$ (of height N and base Y_2) and partition $\overline{P}$ such that $G_2'' \sim G_1''$.

---

*In this part we will need quantities analogous to u, δ, $\bar{\delta}$, N, M, J, G_1, G_2, etc. Although these will be different we will use the same letters. However, ε, K, and L will retain their old meaning.

Let $\hat{P}$ be a partition such that the points in Y_1 have only a finite number of different continuous $\hat{P}$-names of length N, and (a) $|\hat{P} - P| < \frac{1}{20K}\bar{\delta}$.

Because $G_2'' \sim G_1''$, we can find $\hat{\bar{P}}$ such that the points in Y_2 have only a finite number of different continuous $\hat{\bar{P}}$-names of length N (b) $|\hat{\bar{P}} - \bar{P}| < \frac{1}{20K}\bar{\delta}$, and if we form G_1' and G_2' by using $\hat{P}$ and $\hat{\bar{P}}$ instead of P and $\bar{P}$, then (c) $G_1' \sim G_2'$. Let $G_2 = \bigcup_{0\le t\le N} S_t Y_2, \hat{\bar{P}} \vee Q$. Define Q' so that if $G_1 = \bigcup_{0\le t\le N} S_t Y_1, \hat{P} \vee Q'$, then $G_2 \sim G_1$.

As before, it does not matter how we define Q' on $X - \bigcup_{0\le t\le N} Y_1$.

(6) Since $|\bar{P} - \hat{\bar{P}}| < \frac{\bar{\delta}}{10} < \frac{\varepsilon}{10}$, (4.1) and (4.2) imply $\bigvee_{-K_2}^{K_2} S_1^i Q \overset{\frac{2}{10}\varepsilon}{\supset} \hat{\bar{P}}$ and $\bigvee_{-K_3}^{K_3} S_1^i Q \overset{\frac{2}{10}\bar{\delta}}{\supset} \hat{\bar{P}}$. Since $G_1 \sim G_2$ filling most of X and $K_2 < K_3 < \frac{\bar{\delta}}{100}N$, we get that $\bigvee_{-K_2}^{K_2} S_1^i Q' \overset{\frac{3}{10}\varepsilon}{\supset} \hat{P}$ and $\bigvee_{-K_3}^{K_3} S_1^i Q' \overset{\frac{3}{10}\bar{\delta}}{\supset} \hat{P}$. Again, because $|\hat{P} - P| < \frac{\bar{\delta}}{10} < \frac{\varepsilon}{10}$, we have (6.1) $\bigvee_{-K_2}^{K_2} S_1^i Q' \overset{\frac{4}{10}\varepsilon}{\supset} P$, (6.2) $\bigvee_{-K_3}^{K_3} S_1^i Q' \overset{\frac{4}{10}\bar{\delta}}{\supset} P$.

(6.2) and our choice of $\bar{\delta}$ imply that $\left|E(Q', S_{\frac{1}{M}}) - E(Q, S_{\frac{1}{M}})\right| < \delta$. Because $G_1 \sim G_2$, filling most of X, and Q is $\frac{\varepsilon}{100K_2}$ good rel $\frac{1}{M}$, we get that Q' is $\frac{2\varepsilon}{100K_2}$ good rel $\frac{1}{M}$. Hence, by our choice of u, M, δ we get a $\bar{Q}$ satisfying (a) $|\bar{Q} - Q'| < \frac{\varepsilon}{10K_2}$ and (b) $\bar{Q}, S_t \sim Q, S_t$.

Because (a) and (6.1), we get (c) $\bigvee_{-K_2}^{K_2} S_1^i \bar{Q} \overset{\varepsilon}{\supset} P$.

Because $|\bar{P} - \hat{\bar{P}}| < \frac{\bar{\delta}}{10K} < \frac{\varepsilon}{10K}$ and because $\left| L\left(\bigvee_{-K}^{K} S_1^i \bar{P}\right) - Q \right| < \frac{1}{10}\varepsilon$

(see (A)), we get $\left| L\left(\bigvee_{-K}^{K} S_1^i \hat{\overline{P}}\right) - Q \right| < \frac{2}{10}\varepsilon$. Because $G_1 \sim G_2$, we get $\left| L\left(\bigvee_{-K}^{K} S_1^i \hat{P}\right) - Q' \right| < \frac{3}{10}\varepsilon$ and because $|Q' - \overline{Q}| < \frac{\varepsilon}{10}$ and $|P - \hat{P}| < \frac{\overline{\delta}}{10K}$, we get (d) $\left| L\left(\bigvee_{-K}^{K} S_1^i P\right) - \overline{Q} \right| < \frac{5}{10}\varepsilon$. However, (e) $\left| L\left(\bigvee_{-K}^{K} S_1^i P\right) - Q \right| < \frac{1}{2}\varepsilon$ (see (A)).

Because of (d) and (e), we have (f) $|\overline{Q} - Q| < \varepsilon$. Our lemma follows from (b), (c), and (f).

COROLLARY 12. Under the same hypothesis as Lemma 11 we can change conclusion (2) of Lemma 11, to read $\bigvee_{-\infty}^{\infty} S_1^i Q \supset P$.

Proof. The proof is exactly the same as that of Proposition 11, p. 31.

THEOREM 0. Let S_t and $\overline{S}_t$ be Bernoulli flows such that $E(S_1) = E(\overline{S}_1)$. Assume that S_t and $\overline{S}_t$ act on Lebesgue spaces and that the respective σ-algebras separate points. Then S_t and $\overline{S}_t$ are isomorphic.

Proof. Pick finite partitions P and $\overline{Q}$ such that P generates under S_1 and $\overline{Q}$ generates under $\overline{S}_1$. Because of Lemma 2 it will be enough to find Q' such that (1) Q' generates under S_1 and (2) $Q', S_t \sim \overline{Q}, \overline{S}_t$.

We first find a Q such that (a) $Q, S_t \sim \overline{Q}, \overline{S}_t$. (This follows from the Corollary 9.) Because of (a), Lemma 2, and the assumption that $\overline{S}_t$ is a flow built under a function, we get that there is an X_Q satisfying the hypothesis of Lemma 11.

We thus have a P and Q satisfying the conditions of Lemma 11. We can now change Q so that it generates, by Corollary 12.

12. Bernoulli flows of infinite entropy

We say that S_t is a Bernoulli flow of infinite entropy if S_1 is a generalized Bernoulli shift of infinite entropy. We will show that any two such flows are isomorphic.

Recall that an example of such a flow was constructed in [22] (see pages 60 and 61).

LEMMA 1. If S_t is a flow (where $E(S_1) = \infty$), then we can find a good partition P such that $E(P) < \infty$ and P under S_1 generates the sigma algebra generated by P under S_t (that is, we have a factor of finite entropy).

Proof. We must construct a P such that for a.e. x the continuous P-name of x is determined by its P-name under S_1.

Pick $\varepsilon_i > 0$ such that $\sum_1^\infty \varepsilon_i < \infty$, $\varepsilon_{n+1} < \varepsilon_n$. Pick n_i so that

$$\sum_1^\infty - \frac{30}{\varepsilon_k n_k} \log \frac{30}{\varepsilon_k n_k} < \infty, \quad \frac{30}{\varepsilon_i} < n_i, \quad \text{and } \sum \frac{1}{\varepsilon_i n_i} < \infty.$$

(1) Form an R-K set of height n_1 and base Y_1; pick $\frac{1}{\varepsilon_1}$ and numbers, t_i, $0 \le t_i \le \frac{30}{\varepsilon_1} \le \frac{1}{3} n_1$, $20 \le t_{i+1} - t_i \le 30$, $t_0 = 0$ in such a way that every $0 \le x \le 1$ can be approximated to within ε_1 by the fractional part of one of the t_i. Let P_1 be a partition with atoms P^1, P^2, R where $P^1 = \bigcup_{t_i} \left(\bigcup_{t_i \le t \le t_i + 10} S_t Y_1 \right)$,

$P^2 = \bigcup_{n_1 - 10 \le t \le n_1} S_t Y_1$. The P_1-name of x will thus contain groups of $\frac{1}{\varepsilon_1}$ consecutive P^1 intervals of length 10 followed by a P^2 interval. We will define a 1-block in the continuous P_1-name of x to be those t, $t' \le t < t' + n_1$ where $S_{t'}(x) \in Y_1$, partitioned by P^1, P^2, and R. It is clear that the 1-blocks are uniquely determined by the continuous P_1-name of x.

It is easy to see that (a) the P_1-name under S_1 determines each 1-block in the continuous P_1-name up to a translate of ε_1.

(2) At the kth stage we will have a good partition $P_{k-1} = \{P_1 \ldots P^{2k-2}, R\}$ such that the continuous P_{k-1}-name breaks up (in a canonical way) into (k−1)-blocks. Furthermore, the P_{k-1}-name under S_1 determines each (k−1)-block, up to a translate (of the whole block) of $< \varepsilon_{k-1}$.

We will construct P_k as follows: Construct an R-K set of height n_k and base Y_k.

(a) Change P_{k-1} by translating each (k−1)-block in the P_{k-1}-name of each y in Y_k, by $< 2\varepsilon_{k-1}$ so that it begins at a multiple of $2\varepsilon_{k-1}$.

(b) Remove all of the (k−1)-blocks that intersect $\bigcup_{(0\le t\le \frac{30}{\varepsilon_k})} S_t Y_k$ or $\bigcup_{(n_k - 10\le t\le n_k)} S_t Y_k$. (That is, put all of the P^i intervals belonging to these blocks in R.)

(c) Define P^{2k-1} on $\bigcup_{0\le t\le \frac{30}{\varepsilon_k}} S_t Y_k$ as follows: Pick $\frac{1}{\varepsilon_k}$, t_i such that

$20 \le t_{i-1} - t_i \le 30$ and $0 \le t_i \le \frac{30}{\varepsilon_k}$, $t_1 = 0$ and the fractional part of the t_i is ε_k dense in (0, 1). Define P^{2k-1} as $\bigcup_{t_i} \left(\bigcup_{t_i \le t \le t_i + 10} S_t Y_k\right)$.

(d) Define P^{2k} as $\bigcup_{n_k - 10\le t\le n_k} S_t Y_k$.

(3) (a) We now have that the continuous P^k-name breaks up into k-blocks (starting with $\frac{1}{\varepsilon_k}$, P^{2k-1}-intervals and ending with one P^{2k}-interval).

(b) The P_k-name under S_1 determines each k-block up to a translate of $< \varepsilon_k$. (We see this as follows: The P^{2k-1} intervals are determined up to a translate of $< \varepsilon_k$ because of the density of the fractional part of the t_i. Each succeeding (k−1)-block is determined up to a translate of $< \varepsilon_{k-1}$, and since the distance between the beginning of the (k−1)-block and the k-block containing it is a multiple of $2\varepsilon_{k-1}$, this distance is determined exactly by the P_k-name under S_1.)

(4) The partitions P_i converge to a limiting partition P. The reason for this is that the measure of the set E_k, where we changed the partition in going from P_{k-1} to P_k, is so small that $\Sigma\, m(E_k) < \alpha$. (We see this as follows: The part of E_k due to step (a) is $< \frac{1}{5}\varepsilon_{k-1}$ because each P^i-interval has length 10. The part of E_k due to steps (b), (c), and (d) is $< \frac{30}{\varepsilon_k n_k} + \frac{10}{n_k}$.

The ε_k and n_k were picked so that $\sum_1^{\infty} m(E_k) < \infty$.)

(5) $E(P) < \infty$ because $-m(P_{2k-1}) \log [m(P_{2k-1})] \leq -\frac{30}{\varepsilon_k n_k} \log \frac{30}{\varepsilon_k n_k}$ and $-m(P_{2k}) \log[m(P_{2k})] < -\frac{10}{n_k} \log \frac{10}{n_k}$.

(6) The continuous P-name of x is determined by the P-name under S_1. We see this as follows: For each k the continuous P-name contains k-blocks. If we know these k-blocks for all k, then we know the continuous P-name. Each k-block will be contained in a (k+r)-block for any r. The (k+r)-block is determined up to a translate of t_{k+r} by the P-name under S_1, and hence our original k-block is determined up to a translate of $< \varepsilon_{k+r}$ by the P-name under S_1. Thus all of the k-blocks are determined by the P-name under S_1.

LEMMA 2. Let S_t be a flow such that $E(S_1) = \infty$. Then we can find a sequence $\sigma_i \subset \sigma_{i+1}$ of σ-algebras invariant under S_t such that $\bigcup_1^{\infty} \sigma_i$ is the entire σ-algebra and σ_i has the property that there is a good partition $({}^iP, E\,{}^iP) < \infty$ and $\bigvee_{-\infty}^{\infty} S_t({}^iP) = \bigvee_{-\infty}^{\infty} S^j_{\frac{1}{M_i}}({}^iP)$ (for some M_i).

Proof. We will let the P constructed in the previous lemma be ${}^1P$. We will get ${}^2P$ from ${}^1P$ by partitioning ${}^1P^1$ (the first atom of ${}^1P$). We will get ${}^iP$ from ${}^{i-1}P$ by partitioning ${}^1P^i$ (${}^1P^i$ is the ith atom of ${}^1P$ and is an atom of ${}^{i-1}P$). We will do this in such a way that each ${}^1P^i$ interval I in the continuous ${}^{i-1}P$-name of x will be partitioned into a fixed number of intervals, J_ℓ, of equal length (but the J_ℓ will be in different atoms of ${}^iP$ for different I). This will insure that if M_i is an integer and large enough, then the continuous ${}^iP$-name will be determined by the ${}^iP$-name under $S_{\frac{1}{M_i}}$. It is also clear (from the R-K picture) that we can partition the ${}^1P^i$ in such a way that each set in some countable dense class can be approximated arbitrarily well by sets in $\bigvee_{-\infty}^{\infty} S_t({}^iP)$.

LEMMA 3. Let S_t be a flow such that S_1 is a Bernoulli shift of finite entropy. Let H be such that $\bigvee_{-\infty}^{\infty} S_t(H) = \bigvee_{-\infty}^{\infty} S_1^i(H)$ and $E(H,S_1) < E(S_1)$. Then given ε we can find finite partitions $\hat{H}$ and $\hat{R}$ such that (1) $|\hat{H} - H| < \varepsilon$, (2) $\hat{H} \vee \hat{R}$ generate under S_1, (3) any set in $\bigvee_{-\infty}^{\infty} S_t(\hat{H})$ is independent of any set in $\bigvee_{-\infty}^{\infty} S_t\hat{R}$ (we will write this as $\bigvee_{-\infty}^{\infty} S_t(\hat{H}) \perp \bigvee_{-\infty}^{\infty} S_t\hat{R}$), (4) $\hat{H},S_t \sim H,S_t$.

Proof. Pick a Bernoulli flow $\overline{S}_t$ such that $E(\overline{S}_1) = E(S_1)$ and a partition $\overline{H} \vee \overline{R}$, (1) $\overline{H},\overline{S}_t \sim H,S_t$, (2) $\overline{H} \vee \overline{R}$ generate under $\overline{S}$, (3) $\bigvee_{-\infty}^{\infty} \overline{S}_t\overline{H} \perp \bigvee_{-\infty}^{\infty} \overline{S}_t\overline{R}$, (4) $\overline{R}$ is a very good finite partition. (We do this by letting ${}_1\overline{S}_t$ be a Bernoulli flow on ${}_1\overline{X}$ with generator ${}_1\overline{H}$ such that ${}_1\overline{H},{}_1\overline{S}_t \sim H,S_t$. We then take ${}_2\overline{S}_t$ to be a Bernoulli flow on ${}_2\overline{X}$, such that $E({}_1\overline{S}_1) + E({}_2\overline{S}_1) = E(S_1)$, and let ${}_2\overline{R}$ be a good partition that generates under ${}_2S_1$. Let $\overline{S}_t = {}_1\overline{S}_t \times {}_2\overline{S}_t$, $\overline{X} = {}_1\overline{X} \times {}_2\overline{X}$, $\overline{H} = {}_1\overline{H} \times {}_2\overline{X}$, and $\overline{R} = {}_2\overline{R} \times {}_1\overline{X}$.)

Let $P = \overline{H} \vee \overline{R}$.

To prove our lemma we will find $\hat{H}$ and $\hat{R}$ such that (1) $|\hat{H} - H| < \varepsilon$, (2) $(\hat{H} \vee \hat{R},S_t) \sim (\overline{H} \vee \overline{R},\overline{S}_t)$, (3) $\hat{H} \vee \hat{R}$ generate under S_t. Because of Corollary 12, it will be enough to find $\hat{H}$ and $\hat{R}$ satisfy (1) and (2).

We now pick M so that $\frac{1}{100}\varepsilon$ and $\frac{1}{M}$ work for $\overline{H} \vee \overline{R}$ and H. Furthermore, $|H - S_tH| < \frac{1}{100}\varepsilon$ for all t, $|t| < \frac{1}{M}$. Apply Proposition 10 to get δ and u such that if $\tilde{P} = \tilde{H} \vee \tilde{R}$, $|\tilde{H} \vee \tilde{R}| = |\overline{H} \vee \overline{R}|$ and (a) $\tilde{H} \vee \tilde{R}$ is $\frac{1}{5}\varepsilon$ good rel $\frac{1}{M}$, (b) $\bar{d}\left(\left\{S_{\frac{1}{M}}^i \tilde{P}\right\}_1^u, \left\{\overline{S}_{\frac{1}{M}}^i \overline{P}\right\}_1^u\right) < \delta$, (c) $\left|E(\tilde{P},S_{\frac{1}{M}}) - E(\overline{P},\overline{S}_{\frac{1}{M}})\right| < \delta$, (d) $|\tilde{H} - H| < \frac{1}{2}\varepsilon$.

Then there exists $\hat{P} = \hat{H} \vee \hat{R}$ such that $|\hat{H} \vee \hat{R} - \tilde{H} \vee \tilde{R}| < \frac{1}{2}\varepsilon$ and (using (d)) (e) $|H - \hat{H}| < \varepsilon$, (f) $(\hat{H} \vee \hat{R},S_t) \sim (\overline{H} \vee \overline{R},\overline{S}_t)$.

Our problem is thus reduced to finding $\tilde{P}$ satisfying (a), (b), (c), and (d).

(4) Choose a very good finite partition Q such that $|E(Q,S_{\frac{1}{M}}) - E(S_{\frac{1}{M}})| < \frac{1}{100}\delta$. Choose γ so that if $|Q_1| = |Q|$ and $|Q_1 - Q| < \gamma$, then there is a

partition $J, E(J) < \frac{1}{100}\delta$ and $Q_1 \vee J \supset Q$. It will be convenient to assume $\gamma < \frac{1}{100}\delta$ and $\gamma < \frac{1}{100}\varepsilon$.

Choose L to be a multiple of M such that $\frac{1}{L} < \frac{1}{100}\gamma \ell(Q)$.

(5) (a) All but $\frac{1}{100}\gamma$ of the atoms in $\bigvee_{-nL}^{-1} S_{\frac{1}{L}}^{i}(H \vee Q)$ have measure

$$> \frac{1}{2}^{[E(S_{\frac{1}{L}}) + \frac{1}{100L}\delta]nL} = \frac{1}{2}^{[E(S_{\frac{1}{M}}) + \frac{1}{100M}\delta]nM}.$$

Call these good n-$H \vee Q$-atoms.

(b) If n is large enough, all but $\frac{1}{100}\gamma$ of the atoms in $\bigvee_{-nM}^{-1} S_{\frac{1}{M}}^{i}(H)$ have measure $< \frac{1}{2}^{[E(H,S_{\frac{1}{M}}) - \frac{1}{100M}\delta]nM}$. Call these good n-H-atoms.

(c) All but $\frac{1}{100}\delta$ of the atoms in $\bigvee_{-nM}^{-1} \overline{S}_{\frac{1}{M}}^{i}\overline{R}$ have measure $< \frac{1}{2}^{[E(\overline{R},\overline{S}_{\frac{1}{M}}) - \frac{1}{100M}\delta]nM}$. Call these the good n-$\overline{R}$-atoms.

Putting (a), (b), (c), and the fact that $E(H,S_{\frac{1}{M}}) + E(\overline{R},\overline{S}_{\frac{1}{M}}) = E(S_{\frac{1}{M}})$ together, we get for n large enough

(d) If B is a good n-H-atom and A a good n-$H \vee Q$-atom in B and C any n-$\overline{R}$-atom, then $\frac{1}{2}^{\frac{6}{100}\delta n} \cdot m(C) < (\frac{1}{100}\gamma)^2 \frac{m(A)}{m(B)}$.

(6) If n is large enough, we can assume that: (a) If x is in a good n-H-atom (or good n-$\overline{R}$-atom), then the $\frac{1}{M}$ filled H (or $\overline{R}$) name of x is $\frac{1}{20}\varepsilon$ good rel $\frac{1}{M}$.

We get (a) because M was chosen so that H and $\overline{R}$ were $\frac{1}{100}\varepsilon$ good rel $\frac{1}{M}$.

(b) There is a set $E, m(E) > 1 - \frac{1}{100}\gamma$ such that if $x \in E$, tnen there is a set R_x in $-\frac{1}{M} \leq t \leq \frac{1}{M}$, $m(R_x) > \frac{2}{M} - \frac{\varepsilon}{100M}$, and $t \in R_x$ implies that the H-name of x and $S_t(x)$, under S_1 of length nM agree in $> (1 - \frac{1}{10}\varepsilon)nM$ places.

(b) follows from our choice of M and the ergodic theorem.

(7) We now form a continuous R-K set of height n filling all but $\frac{1}{100}\gamma$ of the space. Let F_L be the part of height $\leq \frac{1}{L}$ (that is, if Y is its base then $F_L = \bigcup_{0 \leq t \leq \frac{1}{L}} S_t Y$). Because of the assumption that S_1 is mixing, we can assume that E and the good n-$H \vee Q$-atoms which are contained in good n-H-atoms fill all but $\frac{2}{100}\varepsilon$ of F_L. We will call the part of the n-$H \vee Q$-atoms that lie in F_L, n-$H \vee Q$-F-atoms.

<u>From now on A_j will always refer to a good n-$H \vee Q$-F-atom and B_j to a good n-$H \vee F$-atom.</u>

(8) As before, we will call a set in F_L, Y-measurable if each set of the form $\bigcup_{0 \leq t \leq \frac{1}{L}} S_t(y)$, y in Y, is either in the set or its complement.

We will now modify each A to obtain a Y-measurable set A′. We will do this as follows: Order the good A_i. Suppose we have defined A_j' for $1 \leq i \leq r$ so that the A_j' are disjoint and $\bigcup_1^r A_i' \supset \bigcup_1^r A_i$. We then look at the part $\tilde{A}_{r+1}$ of $A_{r+1} \cap E$ that lies in $F_L - \bigcup_1^r A_i'$, and let A_{r+1}' be the smallest Y-measurable set in $F_L - \bigcup_1^r A_i'$ containing $\tilde{A}_{r+1}$.

B_i' will from now on denote the union of those A_j' such that $A_j \subset B_i$.

(9) (a) We will define $\tilde{Q}$ (as in the finite entropy argument) so that if $y \in Y \cap A_j$, then the continuous $\tilde{Q}$-names of length n of y will be the $\frac{1}{L}$ filled Q-names of any x in A_j. (Thus all x in A_j' have the same $\tilde{Q}$-names under $S_{\frac{1}{L}}$ of length nL and the $S_{\frac{1}{L}}^i A_j'$, $0 \leq i \leq nL$ refine $\tilde{Q}$ on $\bigcup_{i=0}^{nL} S_{\frac{1}{L}}^i (\bigcup A_j')$.)

(b) We will define $\tilde{H}$ so that if $y \in Y \cap A_j'$, then the continuous $\tilde{H}$-names of length n of y will be the same as the $\frac{1}{M}$ filled H-name of any x in A_j.

We thus have that the $\tilde{H}$-names under $S_{\frac{1}{M}}$ of length nM of points in $(\bigcup A_j')$ partition $(\bigcup A_j')$ into the B_i'.

(10) We will now define $\tilde{R}$. We will divide each n-$\overline{R}$-atom into $2^{\frac{6}{100}\delta n}$ equal pieces ($\overline{R}$ pieces of $\bar{r}_j$). We will then pick a B_i', (see (8) for the defini-

tion) and divide it into Y-measurable (see (8)) pieces or B'_i pieces, or simply b'_i, in such a way that there is a 1 - 1 correspondence between the b'_i and the $\overline{R}$ pieces, and the measures of the b'_i are proportional to the measures of the corresponding $\overline{R}$ pieces. We can do this so that

(a) all but $\frac{3}{100}\gamma$ of the b'_i lie entirely in one A'_j and correspond to an $\overline{R}$ piece of a good n-$\overline{R}$-atom. (This follows easily from (5d), which tells us that B'_i cannot contain too many A'_j, and hence that most of B'_i is covered by large A'_j.

(b) We will define $\hat{R}$ so that if $y \in Y \cap b'_i$ and if $\bar{r}_j$ is the $\overline{R}$ piece corresponding to b'_i, then the continuous $\tilde{R}$-name of length n of y will be the same as the $\frac{1}{M}$ filled $\overline{R}$-name of any $\bar{x} \in \bar{r}_j$ (of length n).

(11) $|\tilde{H} - H| < \frac{3}{100}\varepsilon$. We see this as follows: All points in $\bigcup_{i=0}^{\frac{L}{M}} S^i_{\frac{1}{L}} A'_j$ have the same $\tilde{H}$-name under $S_{\frac{1}{M}}$ of length nM, and this name is the n-H-name of the points in A_j. Because of the construction of A'_j, we have that if $x \in \bigcup_{i=0}^{\frac{L}{M}} S^i_{\frac{1}{L}} A'_j$, then the n-$\tilde{H}$-name of x is the same as the n-H-name of $S_t(x) \in E$ for some $-\frac{1}{M} \leq t \leq \frac{1}{M}$. If t^{-1} is in $R_{S_t(x)}$, then the n-H-name of x and $S_t(x)$ agree in $> (1 - \frac{1}{10}\varepsilon)nL$ places (by (6b)). Since E fills all but $(1 - \frac{1}{100}\gamma)$ of $\bigcup A'_j$, we get (11). The same argument as in Lemma 12 shows that $|\tilde{Q} - Q| < \frac{3}{100}\gamma$.

(12) We will check that $\left|E(\tilde{R} \vee \tilde{H}, S_{\frac{1}{M}}) - E(\overline{R} \vee \overline{H}, \overline{S}_{\frac{1}{M}})\right| < \delta$.

(13) (a) Let F'_L be the union of the b'_i (or B'_j pieces) that lie entirely inside some A'_j. Because of (10a), $m(F'_L) > (1 - \frac{3}{100}\gamma)\, m(\bigcup A'_j)$ and hence

$$m\left(\bigcup_{i=0}^{nL} S^i_{\frac{1}{L}}(F'_L)\right) > 1 - \frac{4}{100}\gamma.$$

Because each n-$\overline{R}$-atom was divided into $2^{\frac{6}{100}\delta n}$ $\overline{R}$-pieces, we can find a partition α of F'_L such that $|\alpha| = 2^{\frac{6}{100}\delta n}$ and the partition of F'_L into b'_i is equal to $\alpha \vee \left(\bigvee_{i=0}^{nM} S_{\frac{1}{M}}^{-i}\tilde{R}\right) \vee$ (the partition of F'_L into B'_i).

The above partition of F'_L is the same as $\alpha \vee \left(\bigvee_{i=0}^{nM} S_{\frac{1}{M}}^{-i}(\tilde{R} \vee \tilde{H})\right)$ (cut down to F'_L) by (9b). Since the partition of F'_L into b'_i refines the partition of F'_L into the A'_j we get that

(b) $\alpha \vee \left(\bigvee_{i=0}^{nM} S_{\frac{1}{M}}^{-i}(\tilde{R} \vee \tilde{H})\right)$ partitions F'_L in a finer way than the A'_j.

(c) Because of (b) we easily get that for $0 \leq i < \frac{L}{M}$, $S_{\frac{1}{L}}^{i}\alpha$ together with $\bigvee_{i=0}^{nM} S_{\frac{1}{M}}^{-i}(\tilde{R} \vee \tilde{H})$ partitions $S_{\frac{1}{L}}^{i} F'_L$ in a finer way than the $S_{\frac{1}{L}}^{i} A'_j$.

(d) It is easy to see from (9) that the sets $S_{\frac{1}{M}}^{k}(S_{\frac{1}{L}}^{i} A'_j)$ $0 \leq i < \frac{L}{M}$, $0 \leq k \leq nM$ partition $\bigcup_{i=0}^{nL} S_{\frac{1}{L}}^{i} F'_L$ in a finer way than $\tilde{Q}$.

(e) Because of (c) and (d) we get that if $\bar{\alpha}$ is the partition whose atoms are the $S_{\frac{1}{L}}^{i} F'_L$ divided according to $S_{\frac{1}{L}}^{i}\alpha$ together with the rest of X, then $\bigvee_{-nM}^{nM} S_{\frac{1}{M}}^{i}(\bar{\alpha} \vee \tilde{H} \vee \tilde{R})$ refines $\tilde{Q}$ on $\bigvee_{i=0}^{nL} S_{\frac{1}{L}}^{i} F'_L$. Because of (a) and our choice of Q and γ we get (12).

(14) We will now show that

$$\text{(a) } \bar{d}\left(\left\{S_{\frac{1}{M}}^{i}\tilde{P}\right\}_1^u, \left\{\bar{S}_{\frac{1}{M}}^{i}\bar{P}\right\}_1^u\right) = \bar{d}\left(\left\{S_{\frac{1}{M}}^{i}(\tilde{H} \vee \tilde{R})\right\}_1^u, \left\{\bar{S}_{\frac{1}{M}}^{i}(\bar{H} \vee \bar{R})\right\}_1^u\right) < \delta.$$

(b) It is easy to see from our construction that the distribution of $\tilde{R}$-names under $S_{\frac{1}{M}}$ of length nM is the same in each B'_i and is the same as the distribution of $\bar{R}$-names under $\bar{S}_{\frac{1}{M}}$ in $\bar{X}$ of length nM.

Because of (b) (and the stationary properties of the $\overline{R}$-names in $\overline{X}$) we have that

(c) The distribution of $\tilde{R}$-names under $S_{\frac{1}{M}}$ of length u in each $S^j_{\frac{1}{L}}(B'_i)$, $0 \leq j \leq nL - \frac{uL}{M}$ is the same as the distribution of $\overline{R}$-names under $\overline{S}_{\frac{1}{M}}$ of length u in $\overline{X}$.

Let $\hat{F}_L$ be the union of the B'_i and let $X_1 = \bigcup_{j+1}^{nL - \frac{uL}{M}} S^j \hat{F}_L$. Because all of the points in $S^j_{\frac{1}{L}} B'_i$ have the same $\tilde{H}$-name under $S_{\frac{1}{M}}$ of length u, (c) will imply

(d) on X_1, $\bigvee_0^u S^i_{\frac{1}{M}} \tilde{R} \perp \bigvee_0^u S^i_{\frac{1}{M}} \tilde{H}$ and (e) $\text{dist}\left(\bigvee_0^u S^i_{\frac{1}{M}} \tilde{R} / X_1\right) = \text{dist}\left(\bigvee_0^u \overline{S}^i_{\frac{1}{M}} \overline{R}\right)$.

Since $|\tilde{H} - H| < \frac{1}{10}\delta$ and $m(X - X_1) < \frac{1}{10}\delta$, we have

(f) $\bar{d}\left(\left\{S^i_{\frac{1}{M}} \tilde{H} / X_1\right\}_1^u \left\{\overline{S}^i_{\frac{1}{M}} \overline{H}\right\}_1^u\right) < \frac{1}{10}\delta$. (d), (e), and (f) imply

(g) $\bar{d}\left(\left\{S^i_{\frac{1}{M}} (\tilde{R} \vee \tilde{H}) / X_1\right\}_1^u, \left\{\overline{S}^i_{\frac{1}{M}} (\overline{R} \vee \overline{H})\right\}_1^u\right) < \frac{1}{10}\delta$. Since X_1 is most of X, (g) implies (a).

(15) We must show that $\tilde{H} \vee \tilde{R}$ is $\frac{1}{5}\varepsilon$ good rel $\frac{1}{M}$. It is easy to see that it is enough to show that both $\tilde{H}$ and $\tilde{R}$ are $\frac{1}{10}\varepsilon$ good rel $\frac{1}{M}$.

This is so because for $y \in Y \cap (\bigcup A'_j)$, the continuous $\tilde{H}$- (or $\tilde{R}$-) name of length n is the $\frac{1}{M}$ filled H- (or $\overline{R}$-) name of a point in a good n-H-atom (or an n-$\overline{R}$-atom). Because of (6) these names are $\frac{\varepsilon}{20}$ good rel $\frac{1}{M}$ and they fill X to within $\frac{1}{100}\gamma$.

THEOREM 4. Let S_t and $\overline{S}_t$ be Bernoulli flows of infinite entropy, acting on Lebesgue spaces whose σ-algebras separate points. Then S_t is isomorphic to $\overline{S}_t$.

Proof. We will show that S_t (and $\overline{S}_t$) are isomorphic to the flow $\hat{S}_t$ which is defined as follows. Let T be a Bernoulli shift of infinite entropy acting on Y.

Let f take 2 values β_1 and β_2, β_1/β_2 irrational, and let f be measurable with respect to the independent generator for T. Let $\hat{S}_t$ be the flow with base Y built under the function f.

It is easy to see that if r_i is a monotone sequence increasing to ∞, then we can find σ-algebras $\hat{\sigma}_i$ invariant under $\hat{S}_t$ such that $\hat{S}_t$ restricted to $\hat{\sigma}_i$ is a Bernoulli flow and $E(\hat{S}_1)$, (restricted to $\hat{\sigma}_i$) equals r_i.

Applying Lemma 2 we get that S_t is the increasing union of invariant σ-algebras σ_i. Because there are good generating partitions for the σ_i, we can see by an easy argument that S_t acting on σ_i is a flow (i.e., a flow built under a function). Because factors of Bernoulli shifts of infinite entropy are Bernoulli shifts, we get that S_t acting on σ_i is a Bernoulli flow of finite entropy.

By the isomorphism theorem for Bernoulli flows of finite entropy we can find good finite generators, ${}^1_1P$, for S_t acting on σ_1, and ${}^1_1\hat{P}$, for $\hat{S}_t$ acting on $\hat{\sigma}_1$ and ${}^1_1P, S_t \sim {}^1_1\hat{P},\hat{S}_t$. Pick ε_1. We apply Lemma 3 to σ_2 and $\hat{\sigma}_2$ to get good partitions ${}^1_2P$, ${}^2_2P \sim {}^1_2\hat{P}$, ${}^2_2\hat{P}$ such that (1) $\left|{}^1_2P - {}^1_1P\right| < \varepsilon_1$ $\left|{}^1_2\hat{P} - {}^1_1\hat{P}\right| < \varepsilon_1$, (2) $\bigvee_{-\infty}^{\infty} S_t({}^1_2P) \perp \bigvee_{-\infty}^{\infty} S_t({}^2_2P)$, $\bigvee_{-\infty}^{\infty} \hat{S}_t({}^1_2\hat{P}) \perp \bigvee_{-\infty}^{\infty} \hat{S}_t({}^2_2\hat{P})$, (3) ${}^1_2P,S_t \sim {}^1_2\hat{P},\hat{S}_t \sim {}^1_1P,S_t \sim {}^1_1\hat{P},\hat{S}_t$, (4) ${}^1_2P \vee {}^2_2P$ generate σ_2 and ${}^1_2\hat{P} \vee {}^2_2\hat{P}$ generate $\hat{\sigma}_2$. We can use the isomorphism theorem for finite entropy flows or examine the beginning of the proof of Lemma 3 to conclude that ${}^2_2P$ and ${}^2_2\hat{P}$ could have been chosen so that ${}^2_2P,S_t \sim {}^2_2\hat{P},\hat{S}_t$, and hence (5) ${}^1_2P \vee {}^2_2P,S_t \sim {}^1_2\hat{P} \vee {}^2_2\hat{P},\hat{S}_t$. (The properties that are important to keep in mind are (1), (4), and (5).)

Continuing in this way, we get for each n good partitions ${}^1_nP \ldots {}^n_nP$ and ${}^1_n\hat{P} \ldots {}^n_n\hat{P}$ such that $(\overline{4})$ $\bigvee_1^n {}^i_nP$ and $\bigvee_1^n {}^i_n\hat{P}$ generate σ_n and $\hat{\sigma}_n$,

$(\overline{5})$ $\left(\bigvee_1^n {}^i_nP\right), S_t \sim \left(\bigvee_1^n {}^i_n\hat{P}\right), \hat{S}_t$.

Furthermore, because of (1) and proper choice of ε_1, ε_2, etc., we can assume that ${}_i^jP$ and ${}_i^j\hat{P}$ converge for fixed j as $i \to \infty$. We thus get good partitions ${}^iP$ and ${}^i\hat{P}$ and (6) $\left(\bigvee_1^\infty {}^iP\right), S_t \sim \left(\bigvee_1^\infty {}^i\hat{P}\right), \hat{S}_t$. (If the ${}_i^iP$ were chosen to have one very large atom, then $\bigvee_1^\infty {}^iP$ would be a countable partition.)

We get $\bigvee_1^\infty {}^iP$ (and $\bigvee_1^\infty {}^i\hat{P}$) to generate as follows: Let Q_i be good generators for σ_i. We have that $\bigvee_{-\infty<t<\infty} S_t\left(\bigvee_1^n {}_n^iP\right) \supset Q_n$, and hence

$$\bigvee_{t\in C_n} S_t\left(\bigvee_1^n {}_n^iP\right) \overset{\frac{1}{2}^n}{\supset} Q_n$$

where C_n is a finite collection of numbers. If we choose our $\varepsilon_{n+1} \ldots$, etc. correctly, we get that $\bigvee_{t\in C_n} S_t\left(\bigvee_1^n {}^iP\right) \overset{\frac{1}{2}^{n-1}}{>} Qn.$

PART III

K-automorphisms*

Kolmogorov introduced a class of transformations which we now call K-automorphisms (or K-transformations). This class includes the Bernoulli shifts and is of importance partly because it is a natural class from several points of view and partly because of the large number of examples which are known to be K-automorphisms. It was once thought that the class of K-automorphisms contained nothing more than Bernoulli shifts, and it is the purpose of this section to show that this is false [12].

The definition of K-automorphism that we will work with is the following:

DEFINITION. T is a K-automorphism if there is a finite partition P such that (a) T^iP generates and (b) given K and ε there is an $N > 0$ such that $\bigvee_{-K}^{K} T^iP$ is ε independent of $\bigvee_{N}^{N+m} T^iP$ for all $m > 0$.

Remark. If Q_1 is a refinement of Q and if $\overline{Q}$ is ε^2 independent of Q_1, then $\overline{Q}$ is 2ε-independent of Q. Hence we could replace condition (b) by: Given K and ε there is an $N > 0$ such that $\bigvee_{-K}^{K} T^iP$ is ε-independent of $\bigvee_{\overline{N}}^{\overline{N}+m} T^iP$ for all $\overline{N} > N$ and $m > 0$.

The definition we have just given is not the usual one. It is equivalent to the usual one because of the following easy lemma:

LEMMA. T and P satisfy conditions (a) and (b) in the above definition if and only if (a) T^iP generates, and (b′) the σ-algebra $\bigcap_{n=1}^{\infty} (\bigvee_{n}^{\infty} T^iP)$ contains only sets of measure 0 or 1.

---

*A large part of this section is taken from a paper by P. Shields and Ornstein, "An uncountable family of non-isomorphic K-automorphisms." Many ideas in this section are also due to R. Adler, N. Friedman, and I. Katznelson.

Proof. The "only if" part is trivial and the "if" part follows immediately from the Martingale convergence theorem.

Remark. In the terminology of processes, (a) and (b′) in the above lemma say that the process P,T satisfies the 0 - 1 law.

Part of the interest in K-automorphisms derives from the following theorem:

THEOREM (Rokhlin-Sinai [53, 5, 60]). $\bigcap_{n=1}^{\infty} \bigvee_{i=n}^{\infty} T^i P$ consists of exactly those sets E in $\bigvee_{-\infty}^{\infty} T^i P$ such that the partition consisting of E and its complement has 0-entropy relative to T.

This theorem has several corollaries.

COROLLARY. T is a K-automorphism if and only if $E(P,T) = 0$ for all finite partitions P.

COROLLARY. T is a K-automorphism if and only if $\bigcap_{n=1}^{\infty} \bigvee_{i=n}^{\infty} T^i P$ is trivial for all finite partitions P.

Remark. If $E(P,T) = 0$, then the process P,T is deterministic in the sense that we can predict the future with probability 1 by knowing the past (or even the arbitrarily distant past). This follows from the fact that $E(P,T) = 0$ if and only if $P \subset \bigvee_{-\infty}^{-1} T^i P$.

Thus a mechanical system is K if and only if no deterministic measurements can be made on the system (other than those with probability 0 or 1).

Because of the second corollary a mechanical system is K if and only if any measurement satisfies the 0 - 1 law.

Remark. K-automorphisms are a natural generalization of Bernoulli shifts in the following sense: Bernoulli shifts are those transformations that arise from independent processes. K-automorphisms are those transformations that arise from processes satisfying the 0 - 1 law.

Remark. Much of the interest in K-automorphisms comes from the examples that are known to be K. Sinai showed that the hard sphere gas and geodesic flow on a surface of negative curvature are K [32, 33, 57, 58]. Rokhlin and Yuzvinskii [52, 35] showed that any ergodic automorphism of a compact group is K (we have hardly begun to list the examples known to be K). Because there is a K-automorphism that is not Bernoulli, it is of interest to show that specific examples which were known to be K are in fact Bernoulli. Thus, Katznelson's theorem that any ergodic automorphism of an n-dimensional torus is Bernoulli and the theorem of Weiss and Ornstein that geodesic flow on a surface of negative curvature is Bernoulli definitely give new information. (The same remark holds for theorems like: factors of Bernoulli shifts are Bernoulli.)

The rest of this part will be devoted to a proof of:

THEOREM. There is a K-automorphism that is not Bernoulli.

Outline of construction of a K-automorphism that is not Bernoulli

T and a four set partition $P = \{P_e, P_f, P_s, P_0\}$ will be constructed in stages; at each stage, the definition of T and P will be extended to a larger part of the measure space. The construction can be modified so that P is a generator. Rather than do this we shall restrict T to the σ-algebra $\Sigma(T,P)$. In other words, T can be considered as a shift on a space X of bilateral sequences of the four symbols 0,e,f,s relative to a measure on X (the shift on P-names), our construction merely showing how to determine the measure. Before giving the construction let us informally describe a typical sequence of X.

Each sequence ω in X is made up of nested blocks, called n-blocks. Each n-block has length h(n) and is made up of (n - 1)-blocks. A 1-block begins with a string of f,$\bar{f}_\omega(1)$ units long followed by a string of 0,$\bar{h}(1) = h(1) - f(1)$ units long, and finally a string of e,$f(1) - \bar{f}_\omega(1)$ units long, where $\bar{f}_\omega(1)$ is a uniformly distributed random integer satisfying $1 \leq \bar{f}_\omega(1) \leq f(1) - 1$.

An n-block begins with a string of f,$\bar{f}_\omega(n)$ long, ends with a string of e,$f(n) - \bar{f}_\omega(n)$ long, where $\bar{f}_\omega(n)$ is a random integer satisfying $1 \leq \bar{f}_\omega(n) \leq f(n) - 1$. In between are 2^n $(n-1)$-blocks separated by strings of s as shown here:

$$f\cdots f\ s\cdots s\{(n-1)\text{-blocks}\}\cdots s\{(n-1)\text{-blocks}\}s\cdots s\cdots s\cdots s\{(n-1)\text{-block}\}s\cdots s\ e\cdots$$

The s are put in so that, reading from left to right, we put in strings $i \times s(n)$ units long, $i = 1, 2, \ldots, 2^n + 1$. Thus an n-block begins with an f-string $\bar{f}_\omega(n)$ units long, then has an s-string s(n) units long, then an $(n-1)$-block, then a string 2s(n) units long, then an $(n-1)$-block, ..., then a string $2^n s(n)$ units long, then the final $(n-1)$-block, then a final s-string $2^{n+1}s(n)$ units long, then an e-string $f(n) - \bar{f}_\omega(n)$ units long.

DEFINITIONS. The transformations to be discussed in this paper were informally described in section 1 as shifts on sequences of 0, e, f, and s. This informal description served in fact as the guide to our later proofs, which only make use of the fact that we can indeed specify a consistent set of joint distribution on the set of sequences, thereby making the shift into a measure-preserving transformation. This can be done directly by making use of the block structure (for example, the "f" is inserted independently, and n-blocks are independent of m-blocks for $m \neq n$). We shall, however, give a geometric description of the desired transformation and its shift invariant measure. This will provide a rigorous framework for our later proofs.

Suppose T is a transformation and B is a set such that T^iB, $0 \leq i \leq r-1$ are mutually disjoint. If P is a partition of $\bigvee_{i=0}^{r-1} T^iB$, then we have a <u>gadget</u> with <u>base</u> B, top $T^{r-1}B$, and height r.

In our situation P will have four sets P_0, P_e, P_f, P_s, so the P-name of x will be the sequence $\{x_i\}$ where x_i is 0, e, f, or s, according to whether T^i_x

is in P_0, P_e, P_f, or P_s. If x is in a gadget G, then only some of these will be defined; we will call these the name of x in G.

Two gadgets of the same height $G = (T^iB, 0 \leq i \leq r-1, P)$, and $\overline{G} = (\overline{T}^i\overline{B}, 0 \leq i \leq r-1, \overline{P})$ are isomorphic if corresponding sets in the two partitions $\left(\bigvee_0^{r-1} T^{-i}P\right) \cap B$ and $\left(\bigvee_0^{r-1} \overline{T}^{-i}\overline{P}\right) \cap \overline{B}$ have the same normalized measure. (The normalized measure on A, $\mu(A) > 0$ is the measure $\mu_A(C) = \mu(C \cap A)/\mu(A)$.)

If G is a gadget, a column is defined as follows: Partition the base according to the name (in G) of points in the base and let $\tilde{B}$ be a set in this partition (that is, $\tilde{B}$ is a member of $\left(\bigvee_0^{r-1} T^{-i}P\right) \cap B$, if G is $(T^iB,$ $0 \leq i \leq r-1, P)$). The set $\bigcup_{i=0}^{r-1} T^i\tilde{B}$ is a column.

Given two gadgets G_1, G_2, each partitioned by P and with bases of the same measure, a new gadget $G_1 * G_2$ is defined by independent cutting and stacking. More precisely, suppose for j = 1,2

$$G_j = (T^iB_j; 0 \leq i \leq r_j - 1, P).$$

Partition G_1 according to columns, and let $\{T^iB_{1\ell}\}$ be such a column. Partition B_2 into sets $B_{2\ell}$ such that $m(B_{2\ell}) = m(B_{1\ell})$ and (1) G_2 is isomorphic to $(T^iB_{2\ell}, 0 \leq i \leq r_2 - 1, P)$. Choose any isomorphism of $T^{r_1-1}B_{1\ell}$ onto $B_{2\ell}$ and call it T. This defines a map T from $T^{r_1-1}B_1$ onto B_2 and gives us the gadget (2) $G_1 * G_2 = (T^iB_1; 0 \leq i \leq r_1 + r_2 - 1, P)$.

Details of Construction

$\overline{h}(1)$ and the sequences f(n) and s(n) will be prescribed later. At this point we merely assume that $f(n) \geq 2$ and $s(n) \geq 1$ for all n. Two sequences of gadgets $\overline{G}(n)$ and G(n) will be constructed inductively, which will define T. (The heights of $\overline{G}(n)$ and G(n) will be denoted by $\overline{h}(n)$ and h(n).)

The first stages $\overline{G}(1)$ and G(1) are described as follows: $\overline{G}(1) = (T^i\overline{B}_1,$ $0 \leq i \leq \overline{h}(1) - 1, P)$ where $\bigcup T^i\overline{B}_1 \subseteq P_0$. ($\overline{h}(1)$ is a number to be prescribed later.) That is, $\overline{G}(1)$ is simply a stack of $\overline{h}(1)$ intervals in P_0 which T maps upwards. To construct G(1) we divide $\overline{B}_1$ into f(1) − 1 disjoint sets of equal measure B'_i, $1 \leq i \leq f(n) - 1$, then for each i select f(1) disjoint sets $B_{i,j}$, $1 \leq j \leq f(1)$, $1 \leq i \leq f(1) - 1$ which do not meet $\cup\, T^i \overline{B}_1$ and satisfy $m(B_{i,j}) = m(B'_i)$. Define T on these (except for $B_{i,f(1)}$) so that $T(B_{i,i}) = B'_i$, $T(T^{\overline{h}(1)-1} B_i) = B_{i,i+1}$, $T(B_{i,j}) = B_{i,j+1}$, if $j \neq i$ and $j \neq f(1)$. Let $B_{i,j}$, $j \leq i$ be in P_f and $B_{i,j}$, $j > i$ be in P_e, $B_1 = \cup\, B_{i,1}$, $h(1) = f(1) + \overline{h}(1)$. $G(1) = (T^i B_1, 0 \leq i \leq h(1) - 1, P)$ (see figure 1).

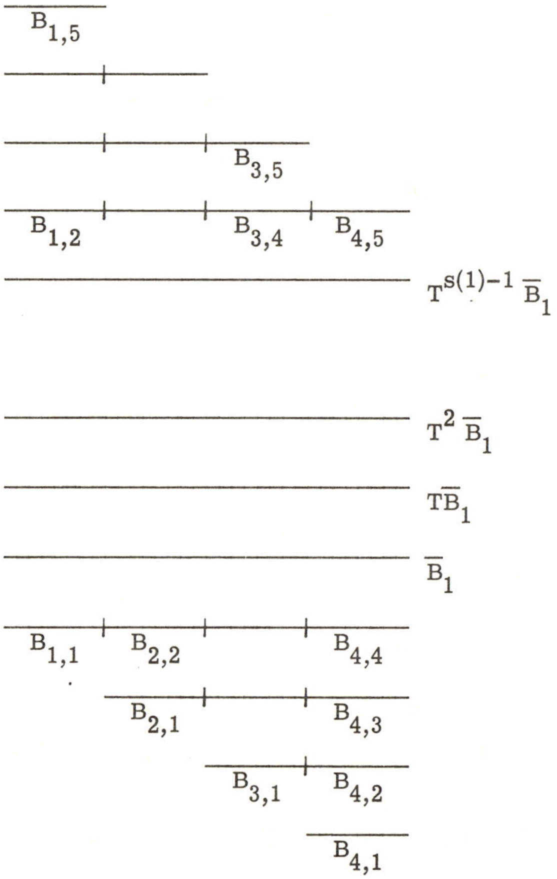

Figure 1

We will now define $G(n)$ and $\overline{G}(n)$ inductively. Assume that $G(n) = (T^i B_n, 0 \le i \le h(n) - 1, P)$ has been constructed. Partition $\overline{B}_n$ into 2^{n+1} disjoint sets of equal measure such that each is the base of a gadget isomorphic to $G(n)$. Call these gadgets $G_1, G_2, \ldots, G_{2^{n+1}}$. (The method of ordering is not important.)

For each i, $1 \le i \le 1 + 2^{n+1}$, another gadget H_i, consisting of sets only in P_s, is constructed by taking i s(n+1) disjoint sets (which do not meet any set in G(n)) $S_{i,j}$, $1 \le j \le i\, s(n+1)$, all having measure equal to the measure of the base of G_1, then defining T so that $T\, S_{i,j} = S_{i,j+1}$, $1 \le j \le i\, s(n+1) - 1$. The gadget H_i is then $\{T^j S_{i,1}, 0 \le j \le i\, s(n+1) - 1, P\}$.

The gadget $\overline{G}(n+1)$ is $H_1 * G_1 * H_2 * G_2 * \ldots * G_{2^{n+1}} * H_{1+2^{n+1}}$ (see figure 2).

To obtain G(n+1), divide the base $\overline{B}_{n+1}$ of $\overline{G}(n+1)$ into f(n+1) − 1 disjoint sets of equal measure $\overline{B}_{n+1,i}$ so that $\overline{G}(n+1)$ is isomorphic to $(T^j \overline{B}_{n+1,i}, 0 \le j \le \overline{h}(n+1) - 1, P)$, where $\overline{h}(n+1)$ is the height of $\overline{G}(n+1)$.

For each $\overline{B}_{n+1,i}$, pick f(n+1) disjoint sets $\overline{B}_{n+1,i,j}$, $1 \le j \le f(n+1)$, which do not meet $\overline{G}(n+1)$ and have measure equal to $\overline{B}_{n+1,i}$. Define T on these (except $\overline{B}_{n+1,i,f(n+1)}$) so that $T\left(\overline{B}_{n+1,i,i}\right) = \overline{B}_{n+1,i}$,

$$T\left(T^{\overline{h}(n+1)-1}\,\overline{B}_{n+1,i}\right) = \overline{B}_{n+1,i,i+1}, \text{ and } T\left(\overline{B}_{n+1,i,j}\right) = \overline{B}_{n+1,i,j+1}, \text{ if } j = i$$

and $j \ne f(n+1)$. Extend P so that $\overline{B}_{n+1,i,j}$ $j \le i$ are in P_f and $\overline{B}_{n+1,i,j}$ $j > i$ are in P_e. Put $B_{n+1} = \bigcup_{i=0}^{f(n+1)-1} \overline{B}_{n+i,i,1}$ and $h(n+1) = \overline{h}(n+1) + f(n+1)$.

G(n+1) is the gadget $(T^i B_{n+1}, 0 \le i \le h(n+1) - i, P)$. (The construction for each $\overline{B}_{n+1,i}$ is similar to that in figure 1.)

Choice of f(n), s(n), and $\overline{h}(1)$

We now wish to define f(n), s(n), and $\overline{h}(1)$ so that the gadgets G(n) have bounded measure, hence interval length can be chosen to obtain a T-invariant probability measure μ on the σ-algebra generated by all the $\{T^i P\}$. Let $f(1) = 2$ (so that there is only one kind of 1-block) and put, for $n > 1$,

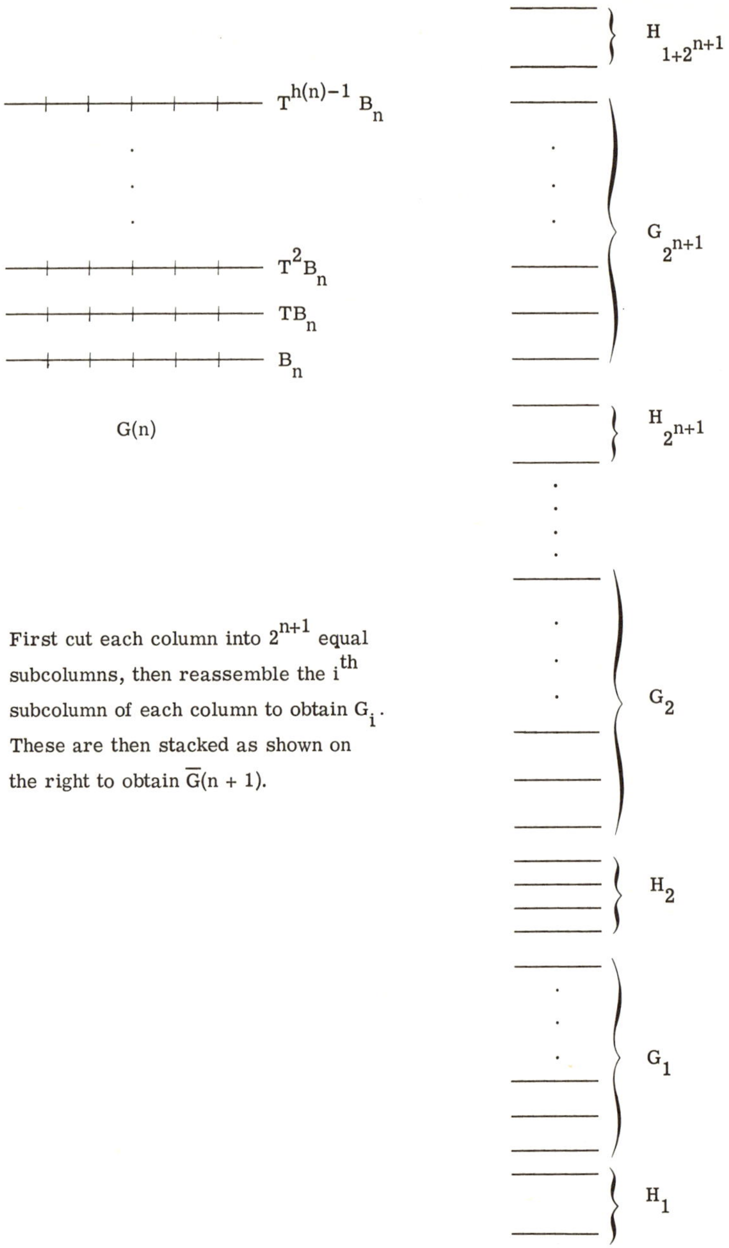

First cut each column into 2^{n+1} equal subcolumns, then reassemble the ith subcolumn of each column to obtain G_i. These are then stacked as shown on the right to obtain $\overline{G}(n+1)$.

Figure 2

$f(n) = n + 1$ and $s(n) = 100n^3$. Certainly, (1) $s(n) > 100 \cdot \sum_{i=1}^{n} f(i)$, $n \geq 1$. Let us show that $\bar{h}(1)$ can be chosen > 200 so that (2) $s(n+1) < (\frac{1}{2})^{10(n+1)} h(n)$, $n > 1$. (Thus, the strings of s separating the n-blocks in an (n+1)-block are short compared to h(n).) To prove this, first note that for $n > 1$, (3) $h(n) = 2^n h(n-1) + (1+2^n)(1+2^{n-1})\, s(n) + f(n)$. It follows that $h(n) > \frac{1}{2} \cdot 2^{n(n+1)/2} \cdot h(1)$. Thus, we need merely choose $\bar{h}(1) < h(1)$ large enough so that $\frac{1}{2} \cdot 2^{n(n+1)/2}(h(1)) \geq 2^{10(n+1)}\, 100(n+1)^3$ holds for $n > 1$ and (1) and (2) will hold.

The total amount of P_s added in going from stage n to n+1 is less than $2^{2(n+2)}\, s(n+1)\, (h(n)^{-1}\, \mu(Gn)) \leq 2^{-8n}\, \mu(G(n))$. Since $f(n) < s(n)$, the total amount of P_f and P_e added at stage n is less than $\leq 2^{-8n}\, \mu(G(n))$. Thus, the $\mu(G(n))$ are uniformly bounded.

Proof that T is a K-automorphism

LEMMA 1. If we are given the name of x we can, for each B_n, determine which $T^i x$ are in B_n.

Proof. Obvious by induction on n.

Remark. We could make our construction on (0,1), and with a little care we could have P together with the B_n generate all measurable sets. Then because of our lemma, $T^i P$ would generate. This, however, will not be necessary.

DEFINITIONS. Let G be a gadget with base B. We define a "rectangle," R, in G as follows: Let E be a set in some $T^i B$, and let $k > i$ be an integer smaller than the height of G. Then $\bigcup_{\ell=0}^{k-i} T^{i+\ell} E$ will be called a rectangle, and E its base. We will say that R is "pure below" if each of the sets $T^{\ell} E$, $-i \leq \ell \leq 0$ is completely contained in some atom of P (which atom depends on ℓ and E).

LEMMA 2. If n is $\leq$ m, then G_m contains disjoint rectangles such that (1) each of these rectangles is isomorphic (as a gadget partitioned by P) to G_n; (2) each rectangle is pure below; (3) the union of these rectangles is equal to $G_n \cap G_m = G_n$, and the union of bases of these rectangles is equal to $F_n \cap G_m = B_n$.

Proof. We will fix n and induct on $m \geq n$. Note that the lemma is obvious for n = m.

We will prove Lemma 2 for $\overline{G}_m$ and G_m, $m > n$. It is obvious that if the lemma holds for $\overline{G}_m$, then it holds for G_m. To show that if the lemma holds for G_m, then it holds for $\overline{G}_{m+1}$, it is enough to show that if the lemma holds for gadget G^1 and G^2, then it holds for $G^1 * G^2$. This is also obvious from the construction.

LEMMA 3. Given an integer ℓ and $\varepsilon > 0$, there exists N such that all $m > N$ have the following property: Let $\{J_i\}$ be the collection of sets of the form $T^j\overline{B}_m$ and $0 \leq j \leq \overline{h}(m)$. Except for a collection $\mathscr{C}$ of J_i, the measure of whose union is less than ε, we have that the distribution of ℓ-names of points in $\bigcup_{j=0}^{f(m)} T^j J_i$ is within ε of the distribution of ℓ-names of points in X.

Proof. Pick n so that the distribution of ℓ-names in G_n is within $\frac{\varepsilon}{100}$ of the distribution of ℓ-names in X and so that $m(X - G_n) < (\frac{1}{100}\varepsilon)^2$.

Pick m so that $h(n)/f(m) < \frac{1}{100}\varepsilon$ and $f(m)/h(m) < \frac{1}{100}\varepsilon$.

It is easy to see that except for $\frac{1}{10}\varepsilon$ of the J_i, $0 \leq i \leq h(m) - f(m)$, all but a set of relative measure $< \frac{1}{10}\varepsilon$ of the points in $\bigcup_{j=0}^{f(m)} T^j J_i$ are in G_n. Because f(m) was large compared to h(n), we get that all but a set of relative measure $< \frac{1}{5}\varepsilon$ of the points in $\bigcup_{j=0}^{f(m)} T^j J_i$ lie in a rectangle (see Lemma 2) which is entirely contained in $\bigcup_{j=0}^{f(m)} T^j J_i$. Lemma 3 is proved because each of these rectangles is isomorphic to G_n.

LEMMA 4. Given an integer ℓ and $\varepsilon > 0$, there exists N such that if $n > N$ we have the following: Let $\{J_i\}$ be the collection T^iB_n. Except for a collection of J_i, the measure of whose union is less than ε, the distribution of the ℓ-names of points in J_i is within ε of the distribution of ℓ-names of points in X.

Proof. Pick n, $n > N$, as in Lemma 3. It is obvious (see figure 1) from the construction of T that B_n will be the union of disjoint sets $B_{n,i,1} \cdots B_{n,f(n),1}$ such that $T^{f(n)-i}B_{n,i,1} = \overline{B}_{n,i}$ is the base of a gadget isomorphic to $\overline{G}_n$. If $f(n) + \ell < r < h(n) - f(n) - \ell$, then

$$T^rB_n = \bigcup_{i=1}^{f(n)} T^rB_{n,i,1} = \bigcup_{i=1}^{f(n)} T^{r-f(n)+i}\overline{B}_{n,i}.$$

We then have that the distribution of ℓ-names of points in T^rB_n is the same as the distribution of ℓ-names of points in $\bigcup_{i=1}^{f(n)} T^i(T^{r-f(n)})\overline{B}_n$, and hence is within ε of the distribution of ℓ-names in X unless $T^{r-f(n)}\overline{B}_n$ is in the exceptional set, $\mathscr{C}$, for Lemma 3. Since ℓ is fixed and since $\frac{f(n)}{h(n)} \to 0$, we get Lemma 4.

THEOREM. T is a K-automorphism.

Proof. We must show the following: Given ℓ, there is an N such that $\bigvee_0^{\ell} T^iP$ is ε-independent of $\bigvee_N^{N+r} T^iP$ for all $r > 0$.

Lemma 4 says that $\bigvee_0^{\ell} T^iP$ is ϵ-independent of Q where Q is the partition of G_n into the T^iB_n. Take $N = h(n)$. For each $m > n$ define Q_m as follows: Lemma 2 gives us a collection of disjoint rectangles R_i in G_m isomorphic to G_n. If F_i is the base of R_i, call T^jF_i, $0 \leq j \leq h(n)$ the levels of R_i. The atoms of Q_m will be the levels of the R_i together with $X - \bigcup R_i$. Lemma 4 says that $\bigvee_0^{\ell} T^iP$ is 2ϵ-independent of Q_m.

Because the R_i are pure below we have that Q_m restricted to

$\left(\bigcup_{i=r+N}^{h(m)} T^i B_m\right)$ refines $\bigvee_N^{N+r} T^i P$ $(N = h(n))$. Thus if n is large enough (making $X - \bigcup R_i$ small), and if h(m) is large enough compared to r, this yields:

$$\bigvee_0^{\ell} T^i P \text{ is } (3\epsilon)^{\frac{1}{2}}\text{-independent of } \bigvee_N^{N+r} T^i P.$$

Rigidity of Blocks

We will now show that the P-names of points have a rigid block structure which will be the key to our later discussion of the isomorphism problem.

An <u>n-block</u> is a sequence $a = \{a_i\}$, h(n) units long of the symbols 0, e, f, s, which appears consecutively in the P-name of some point x in the base B_n, that is, $x \in B_n$ and $T^i x \in P_{a_i}$, $0 \leq i \leq h(n) - 1$. The n-blocks in the P-name of x are uniquely determined by the P-name of x because of Lemma 1 of the previous section.

If $m > n$, then m-<u>order</u> of an n-block a in the name of a point x is i if a is the ith n-block in the unique m-block to which a belongs (reading from left to right). In particular, the $(n + 1)$-order of an n-block is determined once one knows the number of s in front of it.

Let a, b be n-blocks in the P-names of x and y, respectively. The symbol $|a - b|$ will denote the absolute value of the difference of the indices of the first terms of a and b. We will say that a and b are <u>close</u> if

$|a - b| < \sum_{k \leq n} f(k)$. This definition and the properties of f and h described in section 4 immediately imply (1) for any x, y, any n-block a in the P-name of x is close to at most one n-block in the P-name of y.

It is also easy to establish:

LEMMA 1. Let a, b be n-blocks in the P-names of x and y respectively, such that a and b are close and have the same (n+1)-order. Then the (n+1)-blocks a', b' containing a and b are also close and every n-block in a' is close to an n-block in b' of the same (n+1)-order.

Proof. The n-blocks in a (and b) are separated by a number of s-terms depending only on the (n+1)-order of the n-blocks.

The fact that the number of s-terms increases linearly as we move from one n-block to the next within an (n+1)-block provides the key to the next result.

LEMMA 2. Let a be an n-block in the P-name of x. Assume that some (n−1)-block a' in a is close to an (n−1)-block b' in the P-name of some y whose n-order is different from the n-order of a'. Then there is at most one other (n−1)-block a'' in a such that a'' is close to an (n−1)-block b'' in the P-name of y (and b'' cannot lie in b).

Proof. Let b be the n-block in the P-name of y which contains b'. We shall establish (2) no other (n−1)-block in a is close to an (n−1)-block in b.

To prove this, let a'_i and b'_i be the (n−1)-blocks in a and b whose n-orders differ from the n-orders of a' and b' by i. Then $|a'_i - b'_i| \geq i\, s(n) - \sum_{k\leq n-1} f(k)$, and hence (1) of "choice of f(n), s(n), and $\bar{h}(1)$" implies that a'_i is not close to b'_i. Furthermore,

(3) $|a'_i - b'_i| \leq 2^n \cdot 2^n\, s(n) + \sum_{k\leq n-1} f(k) \leq \frac{1}{2}h(n-1)$. We get the first of these inequalities as follows: When i is changed by 1, the left side increases by at most $2^n\, s(n)$, the maximum length of a string of s. Furthermore, $i \leq 2^n$. The second inequality comes from (2) of "choice of f(n), s(n), and $\bar{h}(1)$."

Because of (3), if a'_i is close to any (n−1)-block of b, it must be close to b'_i. This proves (2).

If possible, let $\bar{b}$ be another n-block containing an (n−1)-block $\bar{b}'$ close to an (n−1)-block $\bar{a}'$ in a. From (1) and Lemma 1, $\bar{b}'$ and $\bar{a}'$ have different n-orders, so (2) implies that $\bar{b}$ contains no other (n−1)-block close to an (n−1)-block in a. This proves Lemma 2, for there can be at most two n-blocks containing (n−1)-blocks close to (n−1)-blocks of a.

Most of our subsequent arguments depend upon the fact that if enough of the terms in an n-block are known, the place where the block starts is fairly well determined. A precise form of this is:

LEMMA 3. There is an $\bar{\epsilon} > 0$ such that if a is an n-block in the P-name of x, then for any y either (4) there is an n-block in the P-name of y close to a, or (5) there are more than $\bar{\epsilon}h(n)$ indices i for which $x_i \in a$ and $x_i \neq y_i$.

Proof. Let $h_0(n)$ be the number of 0's in an n-block. (All n-blocks have the same number of 0's.) It is enough to show that the lemma holds if we use $h_0(n)$ instead of h(n) in the statement of (5).

For each n, let ϵ_n be the least non-negative number such that if (4) is not true, then for $x_i \in a$ there are at least $\epsilon_n h_0(n)$ of the indices i for which $x_i \neq y_i$. If (4) does not hold for a given x, y, and n-block a, then at most two (n−1)-blocks in a are close to (n−1)-blocks in the P-name of y (from Lemmas 1 and 2). Since an n-block contains 2^n (n−1)-blocks, we have $\epsilon_n h_0(n) \geq \epsilon_{n-1} (2^n - 2) h_0(n-1) = \epsilon_{n-1} (1 - 2^{-n+1}) 2^n h_0(n-1) =$

$\epsilon_{n-1} (1 - 2^{-n+1}) h_0(n)$,

$$\epsilon_n \geq \epsilon_1 \prod_{j=1}^{n-1} (1 - 2^{-j}).$$

NOTATION. We will refer to any sequence $a = \{a_i\}_0^{\ell-1}$ of 0, e, f, s as a string. We will say that the string occurs in the P-name of x if there are ℓ consecutive terms $b_j \ldots b_{j+\ell}$ in the P-name of x such that $b_{j+i} = a_i$, $0 \leq i \leq \ell - 1$. We will say that the string ε-occurs if $b_{j+i} = a_i$ for more than $(1 - \varepsilon)\ell$ of the i, $0 \leq i \leq \ell - 1$. We will say that the string occurs (or ε-occurs) in an n-block c if the b_{j+i} are in c. If a occurs (or ε-occurs) in an n-block c, we will call the difference between j and the index of the first term in c the position in which it occurs in c.

LEMMA 4. Let a be a string of length ℓ where $\ell > 4h(n-1)$. If a, $\frac{1}{10}\bar{\varepsilon}$-occurs in position i in an n-block c and position j in an n-block d, then $|i - j| < \sum_{i=1}^{n} f(i)$.

Proof. Let $a = \{a_i\}_1^\ell$. We can assume that c (and d) are n-blocks in the P-names of x (and y), and x and y chosen so that $a_i = c_i$ for $> (1 - \frac{1}{10}\bar{\varepsilon})\ell$ of the i, $0 \leq i \leq \ell - 1$ and $c_i \in c$, $0 \leq i \leq \ell - 1$. (We can also assume $a_i = d_i$ for $> (1 - \frac{1}{10}\bar{\varepsilon})\ell$ of the $0 \leq i \leq \ell - 1$ and $d_i \in d$.) Then, by hypothesis, the indices of the first terms of c and d will be $-i$ and $-j$. $\{c_i\}_0^{\ell-1}$ must contain 2 consecutive (n−1)-blocks c' and c'' and less than $\bar{\varepsilon}h(n-1)$ of the i, such that $c_i \in c'$ (or c'') satisfy $c_i \neq d_i$. Applying Lemma 13, we get that there are (n−1)-blocks d' and d'' in d close to c' and c''. By Lemma 2 we have that the n-order c' and d' are the same and by Lemma 1,

$$|i - j| < \sum_{i=1}^{n} f_i .$$

Theorem T is not Bernoulli

Proof. We will suppose that T has an independent generator Q and derive a contradiction.

Pick K such that $\bigvee_{-K}^{K} T^i Q \supset \tilde{P}$, and $|P - \tilde{P}| < \frac{1}{1000}\bar{\varepsilon}$ (this means that the $\tilde{P}$-name of x can be obtained from the Q-name of x by a finite code of length K).

Pick n so that $\frac{h(n-1)}{h(n)} < \frac{1}{100}$ and so that $h(n-1) > 2K$, and $m(X - G_n) < \frac{1}{10}m(X)$. Let $\ell = 5h(n-1)$.

Partition the points according to their Q-names of length 2ℓ. Let c_α be an atom of this partition, the name of whose points is $\alpha = \{\alpha_i\}_0^{2\ell-1}$. Let $c_{\bar{\alpha}}$ be the set of x such that $T^i x \in Q^{\alpha_i}$, $0 \leq i \leq \ell - 1$ and $T^{2s(n)+i} x \in Q^{\alpha_i}$ for $\ell \leq i \leq 2\ell - 1$.

Call x good if (a) the P-name of x of length 3ℓ lies entirely in one n-block and (b) $< \frac{1}{100}\bar{\varepsilon}\ell$ of the i, $0 \leq i \leq 3\ell$ are such that $T^i x$ lies in different P and $\tilde{P}$ atoms.

We will now prove the following: (A) For each α, either all the x in c_α are not good or all the x in $c_{\bar{\alpha}}$ are not good.

We will prove (A) by contradiction. Assume $x \in c_\alpha$ and $y \in c_{\bar{\alpha}}$ are good.

Let β_1 and β_2 be the strings consisting of the terms in the $\tilde{P}$-names of x whose indices range from K to $\ell - 1 - K$, and $\ell + K$ to $2\ell - 1 - K$. Let $\bar{\beta}_1$ and $\bar{\beta}_2$ be the strings consisting of the terms in the $\tilde{P}$-name of y whose indices range from K to $\ell - 1 - K$ and $2s(n) + \ell + K$ to $2s(n) + 2\ell - 1 - K$.

Because of our choice of $\tilde{P}$ and c_α and $c_{\bar{\alpha}}$, we have that $\beta_1 = \bar{\beta}_2$ and $\beta_2 = \bar{\beta}_2$ (since the $\tilde{P}$-name is obtained from the Q-name by a code of length K).

Because x is good, β_1 and β_2, $\frac{1}{20}\bar{\varepsilon}$-occur in an n-block in such a way that their positions differ by ℓ. Because y is good, $\bar{\beta}_1$ and $\bar{\beta}_2$ (and hence β_1 and β_2) $\frac{1}{20}\bar{\varepsilon}$-occur in an n-block in such a way that their positions differ by $2s(n) + \ell$. This contradicts Lemma 4.

We will now show that our theorem follows from (A). Because the T^iQ are independent, we have that $m(c_\alpha) = m(c_{\bar{\alpha}})$. Hence the measure of the set of good x is $\leq \frac{1}{2}$. This, however, is contradicted by our choice of n and the fact that $|P - \tilde{P}| < \frac{1}{1000}\bar{\varepsilon}$.

Further Results

There are several counterexamples that can be obtained by modifying and elaborating the previous construction, and we will now state without proof some of these:

(1) Shields and Ornstein [16] showed that there is an uncountable class of non-isomorphic K-automorphisms of the same entropy. The idea is as follows: In our previous construction the strings of s separating (n−1)-blocks in an n-block were chosen so that their lengths increased as we read from left to right. For each sequence $\{a_n\}_1^\infty$ where $a_n = -1$ or $a_n = 1$, we could define a new transformation $T_{\{a_n\}}$ as follows: at the nth stage of our construction put the strings of s in, in reversed order if $a_n = -1$. It is easy to see that any two of the $T_{\{a_n\}}$ have the same entropy and are isomorphic if the a_n are the same for all but a finite number of n. Furthermore, we can show by a coding argument similar to the one previously used that $T_{\{a_n\}}$ is not isomorphic to $T_{\{b_n\}}$ if $a_n \neq b_n$ for infinitely many n. In particular,

$T_{\{a_n\}}$ is not isomorphic to $T^{-1}_{\{a_n\}}$. The above sheds some light on the complexity of the isomorphism problems for K-automorphisms.

(2) Pinsker conjectured that every transformation was the direct product of a transformation of 0 entropy and a K-automorphism (unless it was already a transformation of 0 entropy or a K-automorphism). This is shown to be false even for mixing transformations [14].

In building the above counterexample we make some constructions which are of interest in their own rights. Here are some of the ideas: In our construction of a K-automorphism which is not Bernoulli (we shall refer to it as T), we got our mixing properties from the random insertion of the f. It is possible to put in the strings of s more carefully—but still deterministically—completely omit the f, and still obtain a mixing transformation (which necessarily has entropy 0). This transformation (we will refer to it as $\overline{T}$) is interesting in its own right in that it has no proper factors and commutes only with its powers. The counterexample to Pinsker's conjecture is obtained as a skew product of $\overline{T}$ and T.

(3) In [13] we construct a K-automorphism with no square root. Jack Clark constructed a K-automorphism with no roots at all [54].

The above questions fall into a general class of questions, namely: which properties of Bernoulli shifts hold for K-automorphisms, or how badly behaved can a K-automorphism be? As noted above, we have some answers, but there is still much work to be done along these lines.

PART IV

Classical Systems*

The purpose of this section is to show how the abstract theory can be applied to classical systems. We will show that (1) ergodic automorphisms of the two-dimensional torus are Bernoulli, and (2) geodesic flow on a surface of negative curvature is a Bernoulli flow.

No attempt is made to give a complete list of applications. For example, Katznelson has shown that every ergodic automorphism of an n-dimensional torus is Bernoullian [50]. (It is not known whether or not every ergodic automorphism of a compact group is Bernoullian, although Rokhlin and Yuzvinskii have shown them to be K [52, 35]). B. Weiss-M. Ratner and L.A. Bunimovitsch have shown that any mixing Anosov flow with a smooth invariant measure is Bernoullian, geodesic flow on a surface of negative being a special case. Sinai has shown that the motion of a billiard ball on a table with a convex obstacle in it and the hard sphere gas confined to a box are K [32, 33]. These are now known to be Bernoullian. The proofs follow the same general outline as the proof we are about to present, but make use of deep results of Sinai. The billiard case was worked out by Gallavotti and Ornstein [64], and Sinai worked out the hard sphere case (oral communication).

Almost nothing is known about the following type of problem: which abstract transformations can be realized as differentiable transformations that preserve a smooth invariant measure (such a system is called classical). For example: Is there a classical K-automorphism (or flow) that is not Bernoulli? Kushnirenko showed that every classical system has finite entropy [34]. We do not know if the topology of the phase space has any

---

*This section is taken from a paper of B. Weiss and D. Ornstein [17].

probabilistic significance for a mechanical system. For example: Which manifolds can support a classical Bernoulli shift (or flow)?

We are reprinting the paper of B. Weiss and D. Ornstein [17] in its entirety, including a few pages of preliminaries which, although redundant here, point out clearly what is being used from the rest of the book.

§1. Approximate independence and Bernoulli shifts

In this section (X, B, μ) is a fixed measure space, $\mu(X) = 1$, and $\varphi: X \to X$ is an invertible measure-preserving transformation. Partitions of X will be denoted by lower case Greek letters $\alpha, \beta, \gamma, \ldots$. $\alpha = \{A_1, A_2, \ldots A_a\}$, $\beta = \{B_1, B_2, \ldots B_b\}, \ldots$. If every atom of α is a union of atoms of β, then we write $\alpha \subseteq \beta$, and $\alpha \vee \beta$ denotes the least common refinement of α and β, i.e., the partition into sets $A_i \cap B_j$. In case $\{\alpha_i\}$ is an infinite sequence of partitions, $\vee \alpha_i$ denotes the smallest σ-algebra with respect to which all the α_i are measurable. A partition α is said to be a generator (this depends on φ) if $\bigvee_{-\infty}^{\infty} \varphi^n \alpha = \beta$, where $\varphi^n \alpha = \{\varphi^n A_1, \ldots \varphi^n A_a\}$. Partitions will be thought of as coming equipped with an ordering of their atoms, and this is inherited in a lexicographic fashion by $\alpha \vee \beta$.

A partition α is said to be independent for φ if for all choices of i_j, $-n \leq j \leq n$, all n (1) $\mu\left(\bigcap_{j=-n}^{n} \varphi^{-j} A_{i_j}\right) = \prod_{-n}^{n} \mu(\varphi^{-j} A_{i_j})$. If φ has an independent generator, then φ is said to be a Bernoulli shift, or B-shift. Before formulating some weaker notions of independence let's introduce a term for a situation that will frequently arise. If a property P holds for all atoms of a partition α with the possible exception of a set of atoms, the measure of whose union is less than ε, then we shall say that P holds for ε-almost every atom of α, or that for ε-a.e. atom A of a P holds. A partition α is a K-partition if for any $B \in \bigvee_{-\infty}^{\infty} \varphi^{-n} \alpha$, given $\varepsilon > 0$ there is an $N_0 = N(\varepsilon, B)$ such that for all $N' \geq N \geq N_0$ and ε-a.e. atom A of $\bigvee_{N}^{N'} \varphi^k \alpha$ we have (2) $|\mu(A \cap B)/\mu(A) - \mu(B)| < \varepsilon$. Clearly B can be replaced by any finite collection of sets. There is a theorem due to Pinsker-Rohlin-Sinai [51, 53] that says, in part, that if φ has a generator α which is a K-partition, then every partition is a K-partition. In that case we say that φ is a K-automorphism (K for Kolmogorov, who introduced the notion). Not every K-automorphism is a B-shift, (see [12]), but if we strengthen somewhat the approximate independence, we will get a property which does imply that φ is a B-shift. First we need to recall some distance functions on the space of partitions. The usual metric is (3) $d(\alpha,\beta) = \sum_i \mu(A_i \Delta B_i)$, where $A \Delta B$ is the symmetric difference $(A \cap B) \setminus A \cap B$. Given two sequences of partitions $\{\alpha_i\}_1^n$, $\{\beta_i\}_1^n$, possibly on different spaces, we could compare their distributions, and we will write (4) $\{\alpha_i\}_1^n \sim \{\beta_i\}_1^n$, if for all k_i, $1 \leq i \leq n$ (5) $\mu(\bigcap_1^n A_{k_i}^{(i)}) = \nu(\bigcap_1^n B_{k_i}^{(i)})$, where μ, ν are the measures on the spaces X, Y, $\alpha_i = \{A_1^{(i)}, \dots A_{a_i}^{(i)}\}$ are partitions of X, and $\beta_i = \{B_1^{(i)}, \dots B_{b_i}^{(i)}\}$ are partitions of Y. We weaken this notion by writing (6) $\bar{d}(\{\alpha_i\}_1^n, \{\beta_i\}_1^n) \leq \varepsilon$ if there are $\bar{\alpha}_i$, $\bar{\beta}_i$ partitions on the same space such that $\{\alpha_i\}_1^n \sim \{\bar{\alpha}_i\}_1^n$ $\{\beta_i\}_1^n \sim \{\bar{\beta}_i\}_1^n$ and (7) $\frac{1}{n} \sum_i d(\bar{\alpha}_i, \bar{\beta}_i) \leq \varepsilon$. If E is any subset of X, then α/E means the partition of E sets of the form $A \cap E$, $A \in \alpha$, where the measure on E has been renormalized so that E has measure one. A partition α is said to be very weakly Bernoullian (VWB) if for every $\varepsilon > 0$ there is an $N_0 = N_0(\varepsilon)$ such that for all $N' \geq N \geq N_0$ and all $n \geq 0$, ε-a.e. atom A of $\bigvee_N^{N'} \varphi^k \alpha$ satisfies (8) $\bar{d}(\{\varphi^{-i}\alpha\}_1^n, \{\varphi^{-i}\alpha/A\}_1^n) \leq \varepsilon$.

We can now formulate the main tools at our disposal for showing that transformations are B-shifts. They are the following two theorems to be found on pages 44, 53, and 54.

THEOREM A. If α is VWB, then $(X, \bigvee_{-\infty}^{\infty} \varphi^n \alpha, \mu, \varphi)$ is a B-shift.

THEOREM B. If $A_1 \subseteq A_2 \subseteq \dots$ are an increasing sequence of φ-invariant σ-algebras $\bigvee_1^\infty A_n = B$ and for each n (X, A_n, μ, φ) is a B-shift, then (X, B, μ, φ) is a B-shift.

To apply Theorem A we need a method for showing that partitions are close in the $\bar{d}$-metric, and we devote the rest of the section to this point. The $\{\alpha_i\}_1^n$-name of x is the sequence $\ell_i = \ell_i(x)$ determined by (9) $x \in A^{(i)}_{\ell_i}$, $\alpha_i = \{A_1^{(i)}, A_2^{(i)}, \dots A^{(i)}_{a_i}\}$. Let e be defined on the integers by $e(j) = 1$ for $j \neq 0$ and $e(0) = 0$.

LEMMA 1.1. Let $\{\alpha_i\}_1^n$ be partitions of X with name functions $\ell_i(x)$, and $\{\beta_i\}_1^n$ partitions of Y with name functions $m_i(y)$. If there is a measure-preserving mapping θ: X $\to$ Y such that

$$(10)\quad \begin{cases} \frac{1}{n}\sum_1^n e(\ell_i(x) - m_i(\theta x)) \leq \varepsilon, & x \in X \setminus E \\ \mu(E) \leq \varepsilon, \end{cases}$$

then (11) $\bar{d}(\{\alpha_i\}_1^n, \{\beta_i\}_1^n) \leq 4\varepsilon$.

<u>Proof</u>. Let $\bar{\alpha}_i = \theta\alpha_i$, $\bar{\beta}_i = \beta_i$. Since θ is measure-preserving, $\{\theta\alpha_i\}_1^n \sim \{\alpha_i\}_1^n$, in fact it's only this that needs to be used of the measure-preserving properties of θ. Note that for two partitions α, β, if $f(x) = 1$ when the α-name disagrees with the β-name and $f(x) = 0$ when the names agree, then (12) $d(\alpha, \beta) = 2\int f(x)\mu(dx)$. Observe that for $\bar{\alpha}_i$ and $\bar{\beta}_i$, (13) $f_i(y) = e(\ell_i(\theta^{-1}y) - m_i(y))$ is 1 if the $\bar{\alpha}_i$-name disagrees with the $\bar{\beta}_i$-name and is zero otherwise. Combining this observation with (12) and (10), we obtain (11).

A version of Lemma 1 will be needed in which the mapping θ is not quite measure-preserving. First a definition. A mapping θ: $X_1 \to X_2$ will be called <u>ε-measure-preserving</u> if there is a set $E_1 \subset X_1$, $\mu_1(E_1) \leq \varepsilon$ and for all $A \subset X_1 \setminus E_1$. (14) $|\mu_2(\theta A)/\mu_1(A) - 1| \leq \varepsilon$. The next lemma is proved by a standard argument and the proof is omitted.

LEMMA 1.2. If $\theta: X_1 \to X_2$ is ε-measure-preserving, and α is a partition of X_1, there is a mapping $\bar{\theta}: X_1 \to X_2$ such that (15) $\{\bar{\theta}\alpha\} \sim \{\alpha\}$ and (16) $\mu_1\{x_1 : \theta x_1 \neq \bar{\theta} x_1\} \leq 3\varepsilon$.

Note that, in general, $\bar{\theta}$ can't be chosen to satisfy (16) and (15) for all α simultaneously. Now for the strengthening of Lemma 1.1.

LEMMA 1.3. Let $\{\alpha_i\}_1^n$ be partitions of X with name functions $\ell_i(x)$, and $\{\beta_i\}_1^n$ partitions of Y with name functions $m_i(y)$. If there is an ε-measure-preserving mapping $\theta: X \to Y$ so that

$$(17)\quad \begin{cases} \dfrac{1}{n}\sum_1^n e(\ell_i(x) - m_i(\theta x)) \leq \varepsilon, \ x \in X \setminus E \\ \mu(E) \leq \varepsilon, \end{cases}$$

then (18) $\bar{d}(\{\alpha_i\}_1^n, \{\beta_i\}_1^n \leq 16\varepsilon$.

Proof. Apply Lemma 1.2 with α replaced by $\bigvee_1^n \alpha_i$. Notice that (17) and (16) imply that (10) holds for $\bar{\theta}$ with ε replaced by 4ε. As we've remarked, the proof of Lemma 1.1 will apply to $\bar{\theta}$ and thus (18) follows.

In applying Lemma 1.3, one of the measure spaces will be a subset of a given measure space, and then ε-measure-preserving should be understood with the measures involved renormalized so that they all have total measure one. The knowledgeable reader will no doubt wonder that no mention has yet been made of entropy. All of the systems that will be discussed have finite entropy, and thus the results on isomorphisms of Bernoulli flows together with Theorem 3.1 imply that for some scaling of the time parameter any two geodesic flows are isomorphic. Indeed, it is shown that if $h(\varphi_1) = h(\tilde{\varphi}_1)$ for two flows and φ_1 and $\tilde{\varphi}_1$ are both B-shifts, then the flows are isomorphic. Since $h(\varphi_t) = |t|\, h(\varphi_1)$, entropy provides the right scaling of the time parameter. To be sure, entropy plays a decisive role in the proofs of Theorems A and B, but no knowledge of it is required in the applications as we shall see below.

§2. Automorphisms of the 2-torus are B-shifts

The method that will be used to prove that certain transformations are B-shifts applies to those oft studied transformations, the automorphisms of the torus. Since the idea of the proof can be seen most clearly in this simple case we devote this section to a proof of

Theorem 2.1. An ergodic automorphism of the 2-torus equipped with Lebesgue measure is a B-shift.

Henceforth, we shall be dealing only with topological spaces and the σ-algebra of measurable sets will be always taken to be the Borel σ-algebra. In [45] it was shown that these automorphisms are Markov shifts while in [50] Theorem 2.1 was proved for the n-torus. The method of proof given below applies to automorphisms of the n-torus that have no repeated eigenvalues of modulus one. For that case it seems that the harmonic analysis of [50] is the most convenient tool. Before coming to the details, let us sketch the idea of the proof. By Theorems A and B of §1 it suffices to show that a refining sequence of partitions of the torus is V.W.B. Let α denote a fixed partition and A an atom of $\bigvee_{N}^{N'} \varphi^k \alpha$. By Lemma 1.3, to see that α is V.W.B. we must map A onto the torus by a mapping θ that is nearly measure-preserving, and that has the additional feature of keeping x and θx close together under the action of φ^k, $k \geq 0$. The construction of θ is done in two steps. First, an auxiliary partition β is chosen which depends on ε. Then the fact that the automorphism is a K-automorphism is used to give a rough, nearly measure-preserving mapping from A onto atoms of β by mapping $A \cap B$ onto B. The contracting and expanding foliations are used "locally" to give a nearly measure-preserving mapping θ from $A \cap B$ to B such that $\{\varphi^k x\}_1^n$ is close to $\{\varphi^k \theta x\}_1^n$. This is done by keeping θx on the contracting fiber that contains x. After this sketch we turn to the details.

Let X be the 2-torus and μ the usual Lebesgue measure on X. An algebraic automorphism φ of the torus X is defined by an integral unimodular 2×2 matrix Φ. Since Φ is unimodular, it preserves μ, and we have an m.p.t.

(X, μ, φ) which we wish to show is a B-shift. The key to the proof is provided by certain invariant fiberings or foliations which we proceed to describe. The ergodicity of φ means that no eigenvalue of Φ can be a root of unity, which in this case means that Φ has two eigenvalues λ_1, λ_2 with $|\lambda_1| < 1 < |\lambda_2|$. Consider Φ as a linear transformation of R^2, and then project the families of lines in the plane parallel to the characteristic directions of Φ to get two φ-invariant fiberings of the torus, $\mathscr{E}$ and $\mathscr{C}$. Let $\mathscr{E}_x$ denote the line through $x \in X$ that is parallel to the characteristic direction associated with λ_2. Since $|\lambda_2| > 1$, it expands when φ is applied to it. That is to say, if $y_1, y_2 \in \mathscr{E}_x$ are separated by a distance d measured along $\mathscr{E}_x$, φy_1 and φy_2 which lie on $\mathscr{E}_{\varphi x}$ are separated by $|\lambda_2| \cdot d$ along $\mathscr{E}_{\varphi x}$. $\mathscr{C}$ corresponds in a similar way to λ_1. These are just the simplest kind of expanding and contracting foliations for a transformation φ. With $\mathscr{E}$ and $\mathscr{C}$ fixed, a set A will be called a parallelogram if it is connected, its closure equals the closure of its interior points, and for any $x, y \in A$ the connected component of $\mathscr{C}_x \cap A$ that contains x, intersects the connected component of $\mathscr{E}_y \cap A$ that contains y in a unique point $z \in A$.

CONVENTION. If A is a parallelogram then for $x \in A$ $\mathscr{E}_x \cap A$, will denote the connected component of $\mathscr{E}_x \cap A$, that contains x. Similarly for $\mathscr{C}_x \cap A$.

Let $\alpha = \{A_1, \ldots A_a\}$ be a fixed partition of X which we want to show is VWB. The only regularity condition that we shall impose on α is (RA): Each atom is a set with dense interior points and the boundary of A_i consists of a finite number of smooth curves. Suppose that $\varepsilon > 0$ is given and that β is an auxiliary partition which will depend on ε. For the meantime we merely assume that $\beta = \{B_0, B_1, \ldots B_b\}$, where the B_i are parallelograms, for $1 \leq i \leq b$, and the measure of B_0 is small. A set $E \subset X$ is said to intersect B_i in an $\mathscr{E}$-tubular subset of B_i, if $x \in E \cap B_i$ implies that $\mathscr{E}_x \cap B_i \subset E \cap B_i$. Our first observation is that for N sufficiently large, most of an atom A in $\bigvee_{N}^{N'} \varphi^k \alpha$ intersects each B_i in an $\mathscr{E}$-tubular subset. For this it suffices to prove:

LEMMA 1. Suppose that α is a partition satisfying (RA), B is a parallelogram, and $\delta > 0$ is given. There exists an N_1 such that for any $N' > N \geq N_1$ and δ-almost every atom $A \in \bigvee_N^{N'} \varphi^k \alpha$ there is a subset $E \subset A$ with (1) $\mu(E)/\mu(A) > 1 - \delta$, (2) E intersects B in an $\mathscr{E}$-tubular subset.

Proof. By the invariance of the foliations it follows that operating by φ does not affect the tubularity of intersections. Let G_k denote the non-tubular intersections of atoms of $\varphi^k \alpha$ with B. That is to say, G_k is the union over $\phi^k A \in \varphi^k \alpha$, of the non-tubular intersections of B with $\varphi^k A$. The non-tubular intersections consist of all $x \in B \cap \varphi^k A$ for which $\mathscr{E}_x \cap B$ is not contained entirely in $B \cap \varphi^k A$. To estimate the size of G_k, it is convenient to apply φ^{-k}. Since B is a parallelogram, and the foliations $\mathscr{E}$ contracts exponentially with negative powers of φ, any point in a non-tubular intersection of A with $\varphi^{-k} B$ must be within distance d_k of the boundary of A where d_k satisfies (3) $d_k \leq C \cdot |\lambda_2|^{-k}$. From (3) it follows that (4) $\mu(G_k) \leq C \cdot |\lambda_2|^{-k}$, since the boundary of A is smooth. Now let N_1 be chosen large enough so that (5) $\mu(G) \leq \sum_{k=N_1}^{\infty} \mu(G_k) \leq \delta^2$, where $G = \bigcup_{N_1}^{\infty} G_k$. Now for $N' \geq N \geq N_1$, δ-almost every atom of $\bigvee_N^{N'} \varphi^k \alpha$ intersects G in a set of relative measure at most δ, since otherwise adding up over those atoms that intersect G in a set of relative measure greater than δ we would contradict (5). This clearly implies the existence of $E \subset A$ satisfying (1) and (2) for δ-almost every atom of $\bigvee_N^{N'} \varphi^k \alpha$.

LEMMA 2.2. Let $\delta > 0$, α a partition satisfying (RA) and B a parallelogram. Then there is an N_2 such that for any $N' \geq N \geq N_2$, and δ-almost every $A \in \bigvee_N^{N'} \varphi^k \alpha$ we have (6) $|\mu(A \cap B)/\mu(A) - \mu(B)| \leq \delta$.

Proof. This follows immediately from the fact that φ is a K-automorphism, a result due to V. A. Rohlin [52]. It can also be obtained, in this case, by using the irrationality of the eigenvalues and Kronecker's theorem on uniform distribution.

These first two lemmas enable us to reduce the problem of mapping atoms of $\bigvee_{N}^{N'} \varphi^k \alpha$ onto X in a way which keeps positive orbits of nearby points close together, to a local problem of mapping $\mathscr{E}$-tubular subsets of parallelograms in such a fashion. It is the generalization of this next lemma that causes the difficulties in the generalizations and so in spite of its simplicity in this setting we call it:

MAIN LEMMA. Given $\delta_1 > 0$ there is a $\delta_2 > 0$ such that if B is a parallelogram of diameter less than δ_2, and E is an $\mathscr{E}$-tubular subset of B, there is a one-to-one, onto mapping θ: E → B such that (7) θ is measure-preserving, and (8) $d(\varphi^k \theta x, \varphi^k x) < \delta_1$, for all $k \geq 0$, $x \in E$.

<u>Proof.</u> If δ_2 is small enough, then (8) will be satisfied provided that $\theta x \in \mathscr{C}_x \cap B$, since distances contract as φ operates on $\mathscr{C}$. Let x_0 be some fixed point in the interior of B, and define θ on $E \cap \mathscr{C}_{x_0} \cap B$ as some one-to-one mapping of $E \cap \mathscr{C}_{x_0} \cap B$ onto $\mathscr{C}_{x_0} \cap B$ which preserves linear Lebesgue measure. There is a natural one-to-one mapping $\pi_{x_0,x}$ between $\mathscr{C}x_0 \cap B$ and $\mathscr{C}_x \cap B$ for any $x \in B$ which satisfies $\pi_{x_0,x}(y) \in (\mathscr{E}_y \cap B) \cap (\mathscr{C}_x \cap B)$ and which preserves the linear Lebesgue measures. Use these $\pi_{x_0,x}$ to define θ on $\mathscr{C}_x \cap B$ by $\pi_{x_0,x} \theta \pi_{x_0,x}^{-1}$. This is possible since E is a tubular set. By Fubini's theorem, the mapping θ so defined on $E \cap B$ is measure-preserving and so (7) holds. As we have remarked, (8) holds if δ_2 is small enough, where this choice clearly depends only on δ_1 and the eigenvalue λ_1.

LEMMA 2.3. Let $\varepsilon, \varepsilon' > 0$ and α a partition that satisfies (RA). Then there is an N, such that for all $N' \geq N$, and for ε-almost every atom

$A \in \bigvee_{N}^{N'} \varphi^k\alpha$ there is a set $E \subset A$, and a one-to-one mapping θ of E onto X such that (9) $\mu(E)/\mu(A) > 1 - \varepsilon$, (10) θ is ε-measure-preserving, (11) for any $k \geq 0$, $x \in E$, $d(\varphi^k x, \varphi^k \theta x) \leq \varepsilon'$.

Proof. Apply the Main Lemma with δ_1 replaced by ε' to obtain δ_2, and let $\beta = \{B_0, B_1, \ldots B_b\}$ be a partition of X such that the B_i are parallelograms, for $1 \leq i \leq b$, with diameter no greater than δ_2, while $\mu(B_0) < \varepsilon/10$. Such partitions clearly exist. Combining Lemmas 2.1 and 2.2 we can find an N such that for ε-a.e. atom A of $\bigvee_{N}^{N'} \varphi^k\alpha$ there is an $E \subset A$, with (9) holding, and for all $B_j \in \beta$, $1 \leq j \leq b$ (12) $|\mu(E \cap B_j)/\mu(E) - \mu(B_j)| < \varepsilon$, and furthermore, E intersects B_j $\mathscr{E}$-tubularly, and $E \cap B_0 = \phi$. Apply the Main Lemma to $E \cap B_j$ for each j separately, $1 \leq j \leq b$, and combine the resulting mappings to a mapping $\theta : E \to X$.

The Main Lemma implies that (11) holds while (10) follows from (12) and the fact that θ is measure-preserving on each $E \cap B_j$.

For later generalizations remark that in the proof of this lemma it was not essential that the partial mappings, given by the Main Lemma, were actually measure preserving. It would have been sufficient to know that they were $\varepsilon/10$-measure preserving, for example, if we had, in addition, replaced ε by $\varepsilon/10$ in (12). The next, final lemma will bring us directly to the proof of the theorem.

LEMMA 2.4. Given a partition α that satisfies (RA) and an $\varepsilon > 0$, there is an N such that for all $N' \geq N$ and ε-a.e. atom A of $\bigvee_{N}^{N'} \varphi^k\alpha$ (13) $\bar{d}(\{\varphi^{-i}\alpha | A\}_1^n, \{\varphi^{-i}\alpha\}_1^n) \leq \varepsilon$ for all $n \geq 1$. In other words, any partition that satisfies (RA) is VWB.

Proof. We would like to use Lemma 2.3 in order to apply Lemma 1.3. First, observe that the $\{\varphi^{-1}\alpha\}$-names as well as the $\{\varphi^{-i}\alpha | E\}$-names of a point x are determined by the α-name of $\varphi^i x$. Since the atoms of α have smooth boundaries if ε' is small enough, conclusion (11) of Lemma 2.3

implies hypothesis (17). This is a consequence of the fact that if ε' is small enough, the set of points x,y such that $d(x,y) < \varepsilon'$ and x and y lie in different atoms of α is of arbitrarily small measure, and averaging over $1 \leq j \leq n$ leads to (17) in Lemma 1.3. Now choose N large enough so that Lemma 2.3 holds with ε replaced by $\varepsilon/16$ and ε' small enough to ensure (17) in Lemma 1.3 with $\varepsilon/16$. Extending θ arbitrarily to $A \setminus E$ and applying Lemma 1.3 we get (13).

Proof of Theorem 2.1. Lemma 4 and Theorem A say that if α is any partition that satisfies (RA), φ is a B-shift on $X, \left(\bigvee_{-\infty}^{\infty} \varphi^k \alpha, \mu\right)$, and since one can easily construct an increasing sequence $\alpha_1 \subseteq \alpha_2 \subseteq \ldots$ of such partitions with $\bigvee_1^{\infty} \alpha_i$, the full σ-algebra of Borel sets, Theorem B completes the proof.

§3. Geodesic flows

A classical example which has played a very important role in the development of ergodic theory is the geodesic flow on a surface of negative curvature. Let M be a complete Riemannian manifold and X its unit tangent bundle, that is to say $x \in X$ is a pair (p,u) where p is a point of M and u is a unit tangent vector to M at p. The geodesic flow $\varphi_t: X \to X$ is defined by $\varphi_t(p,u) = (p',u')$ where to obtain (p',j') one constructs the geodesic on M that passes through p in the direction u, and then one moves out along the geodesic for a distance t, measured in arc length to reach the point p';u' is the direction of the tangent to the geodesic so constructed at p'. The completeness of M guarantees that φ_t is well defined for all t. Clearly $\varphi_t \varphi_s = \varphi_{t+s}$, and one easily checks that φ_t preserves the natural volume on X, denoted by μ, so that we have a measure-preserving flow. Substantial progress was made in the study of the ergodic properties of such flows in the 1930's by G. Hedlund, E. Hopf, and others [47, 48, 49]. After a period of quiescence the subject was revived and advanced about a decade ago with the work of Anosov, Arnold, and Sinai. A detailed account of their work may be found in [46], which will be the basic reference for this section. Our goal in this section is to prove that if M is compact, two-dimensional with negative curvature, then φ_t is a Bernoulli flow, that is to say that φ_t is a B-shift for each t. As noted before, the method we are about to describe applies to more general situations, but we restrict ourselves to this case in order to bring out the main ideas as clearly as possible. We shall follow closely the line of proof in §2 and begin by fixing the notation and generalizing the notion of a parallelogram.

Let M be a compact two-dimensional Riemannian manifold of class-C^3 with curvature K that satisfies (1) $K < 0$ throughout M. As above, X will denote the unit tangent bundle of M and $\varphi_t: X \to X$ the geodesic flow. If dA is the areal measure on M, then φ_t preserves dAdu where du is the usual Lebesgue measure on the circle. Denote dAdu by $d\mu$.

The key to the mixing properties of φ_t lies in the existence of expanding and contracting foliations invariant under φ_t. These were used more or less explicitly by Hedlund and Hopf to prove ergodicity, and then were the main tool in the proof of Anosov and Sinai that φ_t is a K-flow. The basic properties of these foliations were known to J. Hadamard and E. Cartan, although not explicitly as foliations. For a brief discussion with references see [46, §2 Example C]. For the properties we shall need of these foliations see [46, §22–23]. There are two foliations of X into differentiable curves, denoted by $\mathscr{E}$ and $\mathscr{C}$, which are mutually transversal and transversal to the flow direction. They are invariant under φ_t, i.e., $\varphi_t \mathscr{E}_x = \mathscr{E}_{\varphi_t x}$, $\varphi_t \mathscr{C}_x = \mathscr{C}_{\varphi_t x}$ and φ_t expands distances along $\mathscr{E}$ exponentially while it contracts distances along $\mathscr{C}$ exponentially as $t \to +\infty$. Naturally, by its very definition φ_t preserves another foliation, namely the foliation of X by the orbits of $\{\varphi_t\}$, and φ_t preserves distances along this foliation. The parallelograms that played a central role in the Main Lemma of §2 are replaced by <u>P-sets</u> which we now define. Start with a small connected segment $\gamma \subset \mathscr{C}_{x_0}$ of the $\mathscr{C}$-foliation, form the surface $\bigcup_{t=0}^{t_0} \varphi_t \gamma = C$, and then move

C along $\mathscr{E}$ to another surface of that type. That is, for each $x \in C$ take a segment of $\mathscr{E}_x$ that starts at x and ends at e(x) in such a way that

(2) $\bigcup_{x \in C} e(x) \subset \bigcup_{t=0}^{t_1} \varphi_t \gamma', \ \gamma' \subset \mathscr{C}_{x'}$ for some x'. A P-set is then a set of the form (3) $B = \bigcup_{x \in C} \mathscr{E}_{x,e(x)}$ where $\mathscr{E}_{x_1,x_2}$ is the segment of $\mathscr{E}_{x_1}$ that connects x_1 to $x_2 \in \mathscr{E}_{x_1}$. We assume that the lengths of $\mathscr{E}_{x,e(x)}$ are bounded away from zero and infinity.

Our first order of business is to establish the analogues of Lemmas 2.1 and 2.2 so as to reduce our problem to a local one. The convention that $\mathscr{E}_x \cap A$ denotes the connected component of $\mathscr{E}_x \cap A$ that contains x will be made as in §2. An <u>$\mathscr{E}$-tubular subset</u> of a P-set will be defined as before, namely, E is called an $\mathscr{E}$-tubular subset of a P-set B if $E \subset B$ and $x \in E$ implies $\mathscr{E}_x \cap B \subset E$. The partitions which we shall show are VWB are supposed to satisfy some mild regularity assumptions such as:

(RA) α is a finite partition each atom of which is connected with dense interior and has a smooth boundary.

For the remainder of this section we fix a time t_0, denote $\varphi_{t_0} = \varphi$, and concentrate on showing that φ is a B-shift by following the pattern of proof in §2.

LEMMA 3.1. Suppose that α is a partition that satisfies (RA), B is a P-set and $\delta > 0$ is given. Then there is an N_1 such that for any $N' \geq N \geq N_1$, for δ-a.e. atom A of $\bigvee_{N}^{N'} \varphi^k \alpha$ there is a subset $E \subset A$ with

(1) $\mu(E)/\mu(A) > 1 - \delta$, (2) $E \cap B$ is an $\mathscr{E}$-tubular subset of B.

<u>Proof.</u> For each k, consider $\varphi^k \alpha$ and let $G_k \subset B$ denote that part of B that is intersected by an atom $A^{(k)}$ of $\varphi^k \alpha$ in a non $\mathscr{E}$-tubular fashion. That is to say, $x \in G_k$ if for some atom $A^{(k)} \in \varphi^k \alpha$ (3) $x \in A^{(k)} \cap B$, and there exists $y \in \mathscr{E}_x \cap B$, $y \notin A^{(k)}$. We wish to estimate the size of G_k, and to this end note that since the foliations are φ-invariant the tubularity of sets is not affected by applying the transformation φ. From the definition (3) we see

that $\varphi^{-k}G_k$ lies in the set of points whose distance to the boundary of some atom of α is no greater than d_k, where d_k is the maximal length of a connected segment of an $\mathscr{E}$-fiber in $\varphi^{-k}B$. Since α satisfies (RA), the measure of this set is thus bounded by a constant times d_k. The exponential contraction of $\mathscr{E}$ as $t \to -\infty$, and the definition of a P-set imply that

(4) $\sum_N^\infty d_k \to 0$ as $N \to \infty$. Since φ is measure preserving we can conclude

that $\sum_N^\infty \mu(G_k)$ tends to zero as N tends to infinity. Choosing N_1 so that

(5) $\sum_{N_1}^\infty \mu(G_k) < \delta^2$, we have that δ-a.e. atom A of $\bigvee_N^{N'} \varphi^k \alpha$, $(N \geq N_1)$ intersects $\bigcup_{N_1}^\infty G_k$ in a set of measure at most $\delta\mu(A)$, letting $E = A \cap \left(X \setminus \bigcup_{N_1}^\infty G_k \right)$ for the good atoms we have that (1) and (2) hold.

LEMMA 2.2. Let α be a partition that satisfies (RA), B a P-set, and suppose that $\delta > 0$ is given. There is an N_2 such that for any $N' \geq N \geq N_2$ and δ-a.e. atom A of $\bigvee_N^{N'} \varphi^k \alpha$ we have (6) $|\mu(B \cap A/\mu(A) - \mu(B)| \leq \delta$.

Proof. D. V. Anosov and Ya. G. Sinai have proved that φ is a K-automorphism [46, §22, and Appendix], whence (6) follows.

We have at one stroke been able to reduce the problem of showing that a partition α that satisfies (RA) is VWB to a local problem involving the local structure of $\mathscr{E}$ and $\mathscr{C}$. One can extract the geometric information that is needed to prove the Main Lemma from the work of Anosov and Sinai on "absolute continuity" of the foliations $\mathscr{E}$ and $\mathscr{C}$, and indeed this approach is necessary for higher dimensional generalizations. However, for dimension 2 we can give a cleaner argument by using the more precise results of E. Hopf. Translating the results of [49, §3, 7] into our terminology leads to the following:

THEOREM H. Local C^1 coordinates can be introduced in X, (α_1, α_2, s)

such that (7) $\varphi_t(\alpha_1,\alpha_2,s) = (\alpha_1,\alpha_2,s+t)$ and the surfaces α_1 = constant, α_2 = constant correspond to the surfaces obtained by taking a segment γ of $\mathscr{E}$, or $\mathscr{C}$ respectively, and forming $\bigcup_{|t|<t_0} \varphi_t\gamma$.

The content of the theorem lies of course in the statement that the coordinates are C^1. We are now in a position to prove

MAIN LEMMA. Given $\varepsilon_1, \delta_1 > 0$, there is a $\delta_2 > 0$ such that if B is a P-set of diameter less than δ_2 and E is an $\mathscr{E}$-tubular subset of B, there is a one-to-one mapping $\theta : E \to B$ such that (8) θ is ε_1-measure-preserving, (9) $d(\varphi^k\theta x, \varphi^k x) < \delta_1$ for all $k \geq 0$, $x \in E$.

Proof. In the coordinates of Theorem H the invariant measure $d\mu$ takes the form (10) $d\mu = p(\alpha_1,\alpha_2)\, d\alpha_1\, d\alpha_2\, dt$ for some smooth function p. First, choose δ_2 small enough so that $p(\alpha_1,d_2)$ is nearly constant on sets of diameter less than δ_2, where nearly constant is relative to ε_1 in the sense that we require for (α_1,α_2), (α_1',α_2') in the same set B
(11) $|p(\alpha_1,\alpha_2)/p(\alpha_1',\alpha_2') - 1| \leq \varepsilon_1/10$. Then choose δ_2 still smaller, if necessary, to ensure that if $d(x,y) < \delta_2$ and $\alpha_2(x) = \alpha_2(y)$, then for all $k \geq 0$
(12) $d(\varphi^k x, \varphi^k y) \leq \delta_1$. The reason that this is possible is that if $\alpha_2(x) = \alpha_2(y)$, then for some t_1 bounded by a constant times $d(x,y)$, $x = (\alpha_1(x), \alpha_2(x), s(x))$ and $y' = (\alpha_1(y), \alpha_2(y), s(y) + t_1)$ lie on the same contracting fiber of $\mathscr{C}$, thus (12) certainly holds with y replaced by y′. But if δ_2 is small enough, taking note of (7), we would then get (12) as written.

Having chosen δ_2, we construct θ in a fashion similar to that used in the proof of the Main Lemma in §2. Let us call an $\mathscr{E}$-tubular subset $E \subset B$, simple if it has constant width in the direction t. For a simple $\mathscr{E}$-tubular subset, $E \subset B$, the planes α_2 = constant intersect E in sections of equal planar measure, and since B itself is a simple $\mathscr{E}$-tubular set, we can map E onto B along planes α_2 = constant in a fashion that preserves the measure $d\alpha_1\, d\alpha_2\, dt$. Now approximate an arbitrary $\mathscr{E}$-tubular subset by simple ones, use (11) to go from $d\alpha_1\, d\alpha_2\, dt$ to $d\mu$, and note that (12) implies (9) if x and θx lie on the same surface, α_2 = constant to complete the proof.

LEMMA 3.3. Let $\varepsilon, \varepsilon' > 0$ and α, a partition satisfying (RA), be given. Then there is an N such that, for all $N' \geq N$ and ε-a.e. atom $A \in \bigvee_{N}^{N'} \varphi^k \alpha$, there is a set $E \subset A$ and a one-to-one mapping $\theta : E \to X$ such that (13) $\mu(E)/\mu(A)\ 1 - \varepsilon$, (14) θ is ε-measure-preserving, and (15) for any $k \geq 0$, $x \in E$, $d(\varphi^k \theta x, \varphi^k x) \leq \varepsilon'$.

Proof. A standard covering argument shows that for any δ, we can find a finite number of disjoint P-sets, $B_1, B_2, \ldots B_b$, of diameter less than δ such that $B_0 = X \setminus \bigcup_1^b B_i$ has measure less than δ. Using the partition $\beta = \{B_0, B_1, \ldots B_b\}$ and the preceding lemmas the proof of Lemma 2.3 can be applied to yield Lemma 3.3.

LEMMA 3.4. Any partition α that satisfies (RA) is VWB for φ.

Proof. The deduction of Lemma 3.4 from Lemma 3.3 is the same as the proof of Lemma 2.4.

THEOREM 3. The geodesic flow on a compact C^3-manifold of negative curvature is isomorphic to a Bernoulli flow.

Proof. One can find a sequence of partitions $\alpha_1 \subset \alpha_2 \subseteq \ldots \subseteq \alpha_i \subseteq \ldots$ that satisfy (RA) and such that $\bigvee_1^\infty \alpha_n$ is the full σ-algebra of Borel sets. Then Lemma 3.4 and Theorems A and B imply that $\varphi = \varphi_{t_0}$ is a Bernoulli shift. Therefore, φ_t is isomorphic to a Bernoulli flow.

APPENDIX A

The purpose of this appendix is to give the exact definitions of the things about which this book is written, such as "Bernoulli shift" and "isomorphism," and to review some elementary measure theory.

By a measure space (X,σ,m) we will mean a set X, a σ-algebra σ of subsets of X and a countably additive measure m, on σ. We will assume that all subsets of sets of measure 0 are in σ (such a σ-algebra is called complete). We will always assume $m(X) = 1$.

We say that two measure spaces X and Y are isomorphic if there are subsets $X_1 \subset X$ and $Y_1 \subset Y$, $m(Y_1) = m(X)$ and $m(Y_1) = m(Y)$ and a 1 - 1 map ψ of X_1 onto Y_1 such that ψ and ψ^{-1} are measure-preserving.

We say that X is a Lebesgue space if X is isomorphic to the unit interval with Lebesgue measure.

It is easy to see that the unit square in the plane with Lebesgue measure or the unit cube in n-space with Lebesgue measure are Lebesgue spaces.

A Lebesgue space of special importance is the following: Let Y be a finite set, $Y = \{y_i\}_{i=1}^{k}$. Assign each y_i a measure $u(y_i) > 0$ so that $\sum_1^k u(y_i) = 1$. Let X be the space of all doubly infinite sequences of the y_i's, $\{y_{\alpha(j)}\}_{-\infty}^{\infty}$. There is a standard way of defining a σ-algebra σ on X and a measure m on σ called "product measure." (We could get the product measure as follows: Call a set, E, in X a cylinder set if E consists of all those sequences having a specified entry for a finite number of specified coordinates. The m-measure of E will be the product of the u-measures of the specified entries. It is easy to see that if we delete a countable number of points from X, then we can make a 1 - 1 map from X onto the unit interval such that cylinder sets correspond to $\frac{1}{2}$-open intervals of the same measure.

[This identification is similar to the decimal expansion of points.] We can use this identification to define our measurable sets and our product measure.)

We could do a similar construction for the case where Y is countable or a Lebesgue space.

A transformation T acting on X is a 1 - 1 map of X onto itself such that T and T^{-1} are measure-preserving. We say $E \subset X$ is invariant under T if $x \in E$ implies T(x), and $T^{-1}(x)$ are in E.

We say that T acting on X is isomorphic to $\overline{T}$ acting on $\overline{X}$ if there are invariants sets $X_1 \subset X$ and $\overline{X}_1 \subset \overline{X}$ such that $m(\overline{X}_1) = m(\overline{X})$ and $m(X_1) = m(X)$, and there is a 1 - 1 map ψ of X_1 onto $\overline{X}_1$ such that ψ and ψ^{-1} are measure-preserving, and if $x \in X_1$, then $\overline{T}(\psi(x)) = \psi(T(x))$.

Let p_i, $1 \leq i \leq k$ be positive numbers such that $\sum_{i=1}^{k} p_i = 1$. We will define the Bernoulli shift $T_{(p_1 \ldots p_k)}$ as follows: Let Y be a set having k elements y_i, and let y_i have measure p_i. Let X be the space of double sequences of the y_i. We will define the σ-algebra and the product measure as above. $T_{(p_1 \ldots p_k)}$ will act on X by shifting each sequence one to the right (thus $T_{(p_1 \ldots p_k)} \{\alpha_i\} = \{\beta_i\}$ where $\beta_i = \alpha_{i-1}$).

It is easy to see that $T_{p_1 \ldots p_k}$ is isomorphic to a Baker's type transformation as described in the introduction.

$T_{p_1 \ldots p_k}$ is exactly the transformation arising from the process that consists of spinning a roulette wheel with k slots of widths p_i, as in the introduction.

Let P_i be a countable collection of finite (or countable) partitions of X. We define $\bigvee_{-\infty}^{\infty} P_i$ to be the collection of sets E with the property that: given ε there is a K and a set E_K such that the measure of the symmetric difference between E and E_K is $< \varepsilon$ (we write this as $|E - \overline{E}| < \varepsilon$) and E_K is in the smallest algebra of sets containing all the atoms of the P_i, $-K \leq i \leq K$

(we write this as $E_K \subset \bigvee_{-K}^{K} P_i$). It is easy to see that $\bigvee_{-\infty}^{\infty} P_i$ is a σ-algebra and contains all sets of measure 0.

We say that the $\underline{P_i}$ generates if $\bigvee_{-\infty}^{\infty} P_i$ is the entire σ-algebra of X.

There is a minor distinction that we should be careful about. Define $\bar{\bigvee}_{-\infty}^{\infty} P_i$ to be the smallest complete σ-algebra $\bar{\sigma}$, containing all the atoms of the P_i ($\bar{\sigma}$ need not contain all sets of measure 0, but if it contains a set E of measure 0, then it contains all subsets of E). It is clear that $\bigvee_{-\infty}^{\infty} P_i \subset \bar{\bigvee}_{-\infty}^{\infty} P_i$.

LEMMA. Let X be a Lebesgue space P_i, $1 \leq i \leq \infty$, finite (or countable) partitions that generate. Then there is a set $X_1 \subset X$, $m(X_1) = m(X)$, and such that on X_1, $\bigvee_{-\infty}^{\infty} P_i = \bar{\bigvee}_{-\infty}^{\infty} P_i$ and the P_i separate points.

Proof. There is a countable number of sets E_i that separate points and such that any set of measure 0 is covered by a collection of the E_i, the measure of whose union is arbitrarily small. We can find sets $F_i, m(F_i) = 0$ such that on $X - F_i$, E_i is in the smallest σ-algebra generated by the P_i. Now let $X_1 = X - \bigcup F_i$.

Remark. If X is Lebesgue, T ergodic, and T^iP generate, then we change P on a set of measure 0 so that $\bigvee_{-\infty}^{\infty} T^iP$ is the entire σ-algebra. (By the previous lemma we get $\bigvee_{-\infty}^{\infty} T^iP$ to be the entire σ-algebra of an invariant set F of measure 0. Put F in P', then $\bigcap_i T^iP' \supset F$ and has measure 0.)

The problem of showing that a transformation T is isomorphic to $T_{(p_1 \dots p_k)}$ is essentially that of finding a partition $P = \{P^i\}_1^k$ $m(P^i) = p_i$ such that T^iP generates and T^iP are independent (that is, $m\left[\bigcap_1^n T^iP^{f(i)}\right] = \prod_1^n m(P^{f(i)})$.) This is a combinatorial problem and has very little to do with the measure space. If our measure spaces are Lebesgue, then the existence of such a P implies isomorphism in the ordinary sense. The rest of this appendix will be devoted to sketching a proof of this reduction.

LEMMA. Let X and $\overline{X}$ be Lebesgue spaces. Let Q_i and $\overline{Q}_i$ be 2 sequences of ordered partitions of X and $\overline{X}$ such that for all n, dist $\bigvee_1^n Q_i$ = dist $\bigvee_1^n \overline{Q}_i$. Assume that $\bigvee_{-\infty}^{\infty} \overline{Q}_i$ is the entire σ-algebra of $\overline{X}$. Then we can find $X_1 \subset X$ and $\overline{X}_1 \subset \overline{X}$, $m(X_1) = m(X)$ and $m(\overline{X}_1) = m(\overline{X})$ and a measure-preserving map ψ of X_1 onto $\overline{X}_1$ such that $\psi^{-1}\overline{Q}_i = Q_i$. If $\bigvee_{-\infty}^{\infty} Q_i$ is the entire σ-algebra of X, then ψ will be 1 - 1 and ψ^{-1} will also be measure-preserving.

Proof. (1) We can remove a set $\overline{F}$ of measure 0 so that on $\overline{X} - \overline{F}$ we have $\bigvee_{-\infty}^{\infty} \overline{Q}_i = \bigvee_{-\infty}^{\infty} \overline{Q}_i$ and $\overline{Q}_i$ separate points. (This follows from the previous lemma.)

(2) Each x in X lies in a nested sequence of the $Q_i^j, Q_i^{x(i)}$. This corresponds to a nested sequence of the $\overline{Q}_i^j, \overline{Q}_i^{x(i)}$ and $\bigcap_{i=1}^{\infty} \overline{Q}_i^{x(i)}$ contains one point, $\bar{x}$, or is empty. $\psi(x)$ will be $\bar{x}$ (if $\bigcap_1^{\infty} Q_i^{x(i)}$ is non-empty) and we will now show that the set of x for which $\psi(x)$ is not defined has measure 0. For each atom $\overline{Q}_i^j$ we will pick a closed set $\overline{E}_i^j \subset \overline{Q}_i^j$ such that $\sum_j \sum_i m\left(\overline{Q}_i^j - \overline{E}_i^j\right) < \varepsilon$. If x is such that $\psi(x)$ is not defined, then $\bigcap_1^{\infty} \overline{Q}_i^{x(i)}$ is empty; hence for some k, $\bigcap_{i=1}^{k} \overline{E}_i^{x(i)}$ is empty, and hence $\bigcap_{i=1}^{k} \overline{Q}_i^{x(i)}$ lies in the complement of $\bigcap_i \left(\bigcup_j E_i^j\right)$. The sets of the form $\bigcap_{i=1}^{k} \overline{Q}_i^{x(i)}$ that lie in the complement of $\bigcap_i \left(\bigcup_j E_i^j\right)$ where k is as small as possible are disjoint, and hence the measure of their union $< \varepsilon$, and because dist $\bigvee_1^n Q_i$ - dist $\bigvee_1^n \overline{Q}_i$ we have that the measure of the union of the corresponding finite intersections of the Q_i^j has measure $< \varepsilon$. Thus the measure of the set of x for which $\bar{x}$ is not defined has measure $< \varepsilon$ for all $\varepsilon > 0$.

(3) A similar argument to that of (2) shows that the set of $\bar{x}$ in $\overline{X}$ for which there is no $\psi^{-1}\bar{x}$ has measure 0.

(4) We can thus find X_1 and $\overline{X}_1$ and a ψ that maps X_1 onto $\overline{X}_1$, and such that $\psi^{-1}\overline{Q}_i = Q_i$. Hence, if E is in $\overline{\bigvee_{-\infty}^{\infty} \overline{Q}i} \left(= \bigvee_{-\infty}^{\infty} \overline{Q}_i \text{ on } \overline{X}_1\right)$, then $\psi^{-1}E$ is in $\bigvee_{-\infty}^{\infty} Q_i$ and $m(E) = m(\psi^{-1}E)$.

(5) If $\bigvee_{-\infty}^{\infty} Q_i$ were the entire σ-algebra of X, then we could have made the reduction in (1) for both x and $\overline{X}$ and ψ would therefore be an isomorphism.

COROLLARY. Let T and $\overline{T}$ be transformations on Lebesgue spaces, X and $\overline{X}$. Let $\overline{P}$ be a finite (or countable) partition that generates under $\overline{T}$. Let P be a partition such that $\overline{P},\overline{T} \sim P,T$. Then we can find invariant sets $X_1 \subset X$ and $\overline{X}_1 \subset \overline{X}$ such that $m(X_1) = m(X)$ and $m(\overline{X}_1) = m(\overline{X})$ and a measure-preserving map ψ from X_1 onto $\overline{X}_1$ such that $\overline{T}(\psi(x)) = \psi(T(x))$ for all x in X_1. If P generates under T then ψ will be 1 - 1 and ψ^{-1} will also be measure-preserving.

Proof. (1) We need only consider the case where $\overline{\bigvee_{-\infty}^{\infty} \overline{T}^i\overline{P}} = \bigvee_{-\infty}^{\infty} \overline{T}^i\overline{P}$ and distinct points have different $\overline{P}$-names under $\overline{T}$. (Otherwise, we could remove an invariant set of measure 0 from $\overline{X}$ and reduce to this case.) (2) We can apply the previous lemma to get $X_1 \subset X$ and $\overline{X}_1 \subset \overline{X}$ and a mapping ψ of X_1 onto $\overline{X}_1$ so that the P-name of x is the same as the $\overline{P}$-name of $\psi(x)$ for all x in X_1. We could assume that if x_1 and x_2 had the same P-name and that if $\psi(x_1)$ is defined, then $\psi(x_2)$ is defined. Let $\overline{X}_2 \subset \overline{X}_1$ be an invariant set (under $\overline{T}$) of full measure. $X_2 = \psi^{-1}\overline{X}_2 \subset X_1$ has full measure, and it is easy to see that X_2 is invariant under T. (If $x \in X_2$, then there is an $y \in X_2$ so that $\psi(y) = \overline{T}\psi(x)$. The P-name of T(x) is the same as the P-name of y—both of these being the same as the $\overline{P}$-name of $\overline{T}\psi(x)$—and hence ψ is defined on T(x), thus $\psi T(x) = \overline{T}\psi(x)$ and $T(x) \in X_2$.)

THEOREM. If X is a Lebesgue space and E_i are a countable number of measurable sets that separate points, then $\bigvee_{-\infty}^{\infty} E_i$ is the full σ-algebra.

The proof of this theorem is very similar to the proof of the lemma we have just proved; but since the theorem is well known, and since we only use it to prove that a special flow is Bernoulli, we omit its proof.

APPENDIX B

In this appendix I will list the previously known results that we will make use of, see [2, 4, 5, and 60].

(1) Let P be a partition whose atoms are P^i.

Then the entropy of P or E(P) is defined to be $-\Sigma\, m(P^i) \log m(P^i)$.

$$(2)^*\quad E(P \vee Q) - E(Q) = -\sum_i m(Q^i)\left(\sum_j \frac{m(P^j \cap Q^i)}{m(Q^i)} \log \frac{m(P^j \cap Q^i)}{m(Q^i)}\right)$$

where the P^j and Q^i are the atoms of P and Q. This formula says that the increase of entropy when we add P to Q can be calculated by first calculating for each Q^i the entropy of P regarded as a partition of Q^i (where the measure of Q^i is normalized to be 1) and then averaging these numbers according to the measure of the Q^i. The proof of the formula is immediate.

(3) $E(P \vee Q) - E(Q) \leq E(P)$ and we have equality if and only if P and Q are independent. This follows from (2) and the convexity of $x \log x$, $0 \leq x \leq 1$. (Let $g(x) = -x \log x$, then $E(P \vee Q) - E(Q) =$

$$\sum_j \left[\sum_i m(Q^i)\, g\, \frac{m(Q^i \cap P^j)}{m(Q^i)}\right] \leq \sum_j g\left(\sum_i m(Q^i \cap P^j)\right) = \sum_j g[m(P^j)] = E(P).$$ We have = if and only if, for each fixed j, $m(Q^i \cap P^j)/m(Q^i)$ is the same for all i.)

(4) Fix the number of atoms in P, then the max of E(P) occurs exactly when all the atoms have the same measure. (This also follows from the convexity of $-x \log x$ [we need only look at 2 atoms of P].)

(5) If Q refines $\overline{Q}$, then $E(P \vee Q) - E(Q) \leq E(P \vee \overline{Q}) - E(\overline{Q})$. (This follows from (2) and (3), or the same argument as used in (3).)

---

*The left side of (2) is sometimes written as E(P/Q).

(6) We define E(P,T) to be

$$\lim_{n\to\infty} \frac{1}{n} E\left(\bigvee_{0}^{n-1} T^i P\right) = \lim_{n\to\infty} \left[E\left(\bigvee_{0}^{n} T^i P\right) - E\left(\bigvee_{0}^{n-1} T^i P\right)\right].$$ (The right side has a limit because it is monotone decreasing by (5).)

(7) $E(P,T) \leq E(P)$. We have = if and only if $T^i P$ are independent. (This follows from (6) and (3).)

(8) Fix k and ε, then there is a δ such that if P and $\overline{P}$ have k atoms and $|P - \overline{P}| < \delta$, then $|E(P,T) - E(\overline{P},T)| < \varepsilon$. (We can write $P \vee \overline{P}$ as $P \vee Q$ or $\overline{P} \vee \overline{Q}$ where $E(Q) < \varepsilon$, $E(\overline{Q}) < \varepsilon$. Then $E(P,T) < E(P \vee \overline{P},T) = E(\overline{P} \vee \overline{Q},T) \leq E(\overline{P},T) + E(\overline{Q})$. The same holds for $\overline{P}$.)

(9) Given k and ε, there is a δ such that if $\bigvee_{-K}^{K} T^i P \supset^{\varepsilon} Q$, $|Q| = k$, then $E(P,T) \geq E(Q,T) - \varepsilon$ (see "notation"). (This follows from (8) and (6), which implies that $E(P,T) = E\left(\bigvee_{-K}^{K} T^i P, T\right)$.)

(10) We define E(T) to be sup E(P,T) where the sup is taken over all finite partitions P.

(11) If P generates under T, then $E(P,T) = E(T)$. (This follows from (9).)

(12) If T is the Bernoulli shift p_i, $i = 1 \ldots k$, then $E(T) = \sum_{i}^{k} p_i \log p_i$. (This follows from (11), (3).)

More detailed proofs of (1)–(12) can be found in Billingsley [2].

We will also need the following theorem due to Shannon and McMillan.

THEOREM (Shannon-McMillan). If T is ergodic and P a finite partition, then given ε we can find an N such that if $n > N$, then there is a collection C of atoms of $\bigvee_{0}^{n-1} T^i P$ such that $m(C) > 1 - \varepsilon$, and if $A \in C$, then the measure of A is between $\frac{1}{2}^{(E(P,T)+\varepsilon)n}$ and $\frac{1}{2}^{(E(P,T)-\varepsilon)n}$. A proof of a sharpening of this result due to Breiman will be found in Billingsley [2]. In the cases where P,T is an independent or multi-step Markov process, the theorem results from a simple application of the ergodic theorem.

THEOREM (Rokhlin-Kakutani). If T is ergodic, then given ε and n there is a set F such that $T^i F$, $0 \leq i \leq n-1$ are disjoint and $\left(m \bigcup_0^{n-1} T^i F\right) >$ $1 - \varepsilon$. (The measure of the whole space is assumed to be 1.) (A proof can be found in [6].)

Remark. The R-K theorem can be strengthened in the following way: Let P be a partition. Then F can be chosen with the additional property that dist (P/F) = dist P. This is obvious if T is mixing, and it remains true if T is only ergodic [5, Theorem 2, section 2]. (Since Bernoulli shifts and K-automorphisms are mixing, there would be very little lost if we assume all transformations in this book to be mixing.)

We will also need the following well known combinatorial fact.

MARRIAGE LEMMA. Suppose we have a collection of n boys and a collection of $m \geq n$ girls. Suppose that for each group of k boys, $1 \leq k \leq n$, they know between them at least k girls. Then each boy can marry a girl that he knows (and no two boys marry the same girl). A proof can be found in [59], or [7, lemma, section 1].

APPENDIX C

After defining the $\overline{d}$ distance between processes, we gave two other definitions and stated without proof that they were equivalent. Since this equivalence was never used in the book, we left its proof (which is relatively routine) to this appendix.

We will now show that the three definitions of $\overline{d}$ are equivalent. For this purpose, we will call definition (1) (on p. 24) $\overline{d}_1$, definition (2) (p. 24) $\overline{d}_2$, and the original definition (p. 23) $\overline{d}$.

$\overline{d}_1 = \overline{d}$: Suppose $\overline{d}((\mathscr{P},T),(\mathscr{Q},S)) = \gamma$. We will show that $\overline{d}_1((\mathscr{P},T),(\mathscr{Q},S)) = \gamma$ (it is obvious that $\overline{d}_1((\mathscr{P},T),(\mathscr{Q},S))$ cannot be $< \gamma$). To do this we will put a shift invariant measure on sequences of pairs $\{\alpha_i,\beta_i\}$ where each α_i and β_i is an integer between 1 and k, and such that the measure induced on the sequences $\{\alpha_i\}$ is the same as that induced by $\mathscr{P}$,T, and the measure induced on the sequences $\{\beta_i\}$ is the same as that induced by $\mathscr{Q}$,S, and furthermore the measure of all sequences such that $\alpha_0 \neq \beta_0$ is $\leq \gamma$. U will be the shift on $\{\alpha_i,\beta_i\}$. $\overline{\mathscr{P}}$ will partition the sequences according to α_0, and $\overline{\mathscr{Q}}$ will partition the sequences according to β_0. (This U may not be ergodic, but we will worry about that later.)

To get the measure on sequences $\{\alpha_i,\beta_i\}$ we will first define, for each n and m a measure on sequences $\{\alpha_i,\beta_i\}_{-m}^{n}$ in such a way that the measure assigned to a particular sequence $\alpha_i,\beta_i, -m \leq i \leq m$ is the same as the measure assigned to the union of all sequences $\{\alpha_i,\beta_i\}_{-\overline{m}}^{\overline{n}}$, $\overline{n} \geq n$, $\overline{m} \geq m$ which agree with our original sequence for $-m \leq i \leq n$. (We will then get a measure on all infinite sequences by a simple and obvious case of the Kolmogorov extension theorem. We could, as in the case of product measure which we sketched in Appendix A, make the cylinder sets correspond to half-open intervals and use this to get our measure.)

We will get the measure on finite sequence as follows: for each ψ, φ, and n, $\{\psi(T^i\mathscr{P})\}_1^n\ \{\varphi(S^i\mathscr{Q})\}_1^n$ gives us a measure on sequences of α_i, β_i of length n. It also gives us a measure on sequence of length r for $r \leq n$. (Simply average the measures given by $\{\psi(T^i\mathscr{P})\}_\ell^{\ell+r}\ \{\varphi(S^i\mathscr{Q})\}_\ell^{\ell+r}$ as ℓ goes from 1 to n−r, giving each of the above measure weight 1/n−r.) We will now choose a sequence $\varphi_{n_i}, \psi_{n_i}, n_i$ such that, for each r, the measure on sequences of length r defined as above by the $\varphi_{n_i}, \psi_{n_i}, n_i$ converges. (This can be done by a standard diagonalization.) Because $\bar{d} = \gamma$, we can require $\lim_{i\to\infty} \frac{1}{n_i} \sum_{j=1}^{n_i} |\psi_{n_i}(T^j(\mathscr{P})) - \varphi_{n_i}(S^j(\mathscr{Q}))| = \gamma$. We now have a measure on sequences of length r for all r, and, in particular, a measure on $\{\alpha_i, \beta_i\}_{-m}^n$. It is easy to see that the consistency conditions are satisfied, and that the measures therefore extend to a measure on all infinite sequences $\{\alpha_i, \beta_i\}$. It is also easy to see that the measure is invariant under the shift U. It is also easy to see that the measure of the set of sequences such that $\alpha_0 \neq \beta_0$ is γ.

<u>We will now show that U could be chosen to be ergodic.</u> We will do this as follows: Each x corresponds (under U and $\overline{\mathscr{P}} \vee \overline{\mathscr{Q}}$) to a doubly infinite sequence $\{{}_x\alpha_i, {}_x\beta_i\}$ (${}_x\alpha_i$ being the atom of $\overline{\mathscr{P}}$ which contains $U^i x$). By the ergodic theorem, for a.e. x each sequence of length r will occur with a limiting frequency (i.e., fix $\{\bar{\alpha}_i, \bar{\beta}_i\}_{i=1}^r$, then the percentage of $\ell \,|\, |\ell| < n$, such that ${}_x\alpha_{\ell+i} = \bar{\alpha}_i$ and ${}_x\beta_{\ell+i} = \bar{\beta}_i$ for $1 \leq i \leq r$, will tend to a limit as n tends to ∞). These frequencies will give a measure on finite sequence which extends (as before) to a shift invariant measure on doubly infinite sequences. Thus, to a.e. x corresponds a measure-preserving transformation U_x acting on X_x, and a generating partition $\overline{\mathscr{P}}_x \vee \overline{\mathscr{Q}}_x$ of X_x which partitions the sequences according to their first coordinate. We will show that U_x is ergodic for a.e. x. We will then be finished, because for a.e. x, $\overline{\mathscr{P}}_x$, U_x will be the same as $\mathscr{P}$,T and $\overline{\mathscr{Q}}_x$, U_x the same as $\mathscr{Q}$,S. The set of x where $|\mathscr{P}_x - \mathscr{Q}_x| > \gamma$ will have zero measure (because $\bar{d} = \gamma$). We could therefore find a U_x with all the desired properties.

LEMMA. U_x is ergodic for a.e. x.

Proof. This follows easily from theorems of Rochlin, but we will sketch a proof for the sake of completeness.

It will simplify notation if we let $\mathscr{R} = \overline{\mathscr{P}} \vee \overline{\mathscr{Q}}$. Then to each x corresponds a sequence which we will call $\{{}_x\xi_i\}_{-\infty}^{\infty}$. The only properties of U that we will use (or that we have) are that U is measure-preserving and invertible.

As we saw before for a.e., x $\{{}_x\xi_i\}$ gives us a measure $u_n(x)$ on finite sequences of length n for all n.

We will say that the $u_n(x)$ has property A if given a sequence $\{\gamma_i\}_1^{\ell}$ and $\varepsilon > 0$, then there is an N such that if $n > N$, then we can find a collection C of sequence of length n of measure $> 1 - \varepsilon$, and such that the frequency of $\{\gamma_i\}_1^{\ell}$ in any two sequences in C differs by less than ε. (By frequency of $\{\gamma_i\}_1^{\ell}$ in $\{\xi_i\}_1^n$ we mean the fraction of ℓ such that $\xi_{\ell+i} = \gamma_i$, $1 \leq i \leq \ell$.)

We will now show that for a.e. x the $u_n(x)$ induced by ${}_x\xi_i$ satisfies property A. We first note that fixing $\{\gamma_i\}_1^{\ell}$ and ε, the space on which U acts breaks up into invariant sets such that if y and $\bar{y}$ belong to the same set, then the frequency of $\{\gamma_i\}_1^{\ell}$ in $\{{}_y\xi_i\}_{-\infty}^{\infty}$ and $\{{}_{\bar{y}}\xi_i\}_{-\infty}^{\infty}$ differ by less than $\frac{1}{10}\varepsilon$. We now can consider these sets separately. Call one of them $\overline{X}$, and call the frequency of $\{\gamma_i\}_1^{\ell}$ of one of the points in $\overline{X}$, f. Applying the ergodic theorem, we can find a set $\overline{X}_1 \subset \overline{X}$, and an N, such that $m(\overline{X}_1) > (1 - \varepsilon^2)\, m(\overline{X})$, and if $y \in \overline{X}_1$ and $n > N$, then the frequency of $\{\gamma_i\}_1^{\ell}$ in $\{{}_y\xi_i\}_1^n$ is within $\frac{1}{2}\varepsilon$ of f. One more application of the ergodic theorem gives us that there is a part B of $\overline{X}$ having measure greater than $(1 - \varepsilon)\, m(\overline{X})$, and such that if $x \in B$, then the frequency that $U^i x$ is in $\overline{X}_1$ tends to a limit which is greater than $1 - \varepsilon$. Thus, for any x in B the measures induced by $\{{}_x\xi_i\}_{-\infty}^{\infty}$ satisfy property A.

We will now show that if the $u_n(x)$ induced by $\{{}_x\xi_i\}_{-\infty}^{\infty}$ satisfies property A, then U_x will be ergodic. We first note that if property A holds for each sequence of length ℓ, it holds for any collection C of strings of length ℓ (we look at the frequency of occurrence of strings in C). We will now argue by contradiction.

Assume U_x is not ergodic. Then there would be a set E and $U_x(E) = E$.

We could then approximate E by a collection C of sequences of length ℓ (that is, we could find a set E', approximating E and consisting of those y in X_x such that $\{{}_y\gamma_i\}_1^\ell$ is in C, where $\{{}_y\gamma_i\}_1^\ell$ is the sequence corresponding to U_x,y and $\mathscr{P}_x \vee \mathscr{Q}_x$ [we could have $|E - E'| < \epsilon^2\, m(E)$ and $|E - E'| < \epsilon^2\, m(X_x - E)$]). Then for n long enough and for most ($> (1 - \epsilon)\, m(E)$) of the y in E, $\{{}_y\gamma_i\}_1^n$ will have a high ($> 1 - \epsilon$) frequency of C, and for most of the y in $(X_x - E)$ $\{{}_y\gamma_i\}_1^n$ will have a low ($> \epsilon$) frequency of C.

$\bar{d} = \bar{d}_2$: We will now show the equivalence of the second and original definitions of $\bar{d}$. Suppose $\bar{d}((\mathscr{P},T),(\mathscr{Q},S)) = \alpha$. We will show that $\bar{d}_2((\mathscr{P},T),(\mathscr{Q},S))$ must be $\geq \alpha - \gamma$ for all $\gamma > 0$. Suppose this were not so for some $\gamma > 0$. For fixed ℓ and $\epsilon > 0$, we could find an N such that if $n > N$, then there are 2 classes C_1^n and C_2^n of sequences of length n having measure $> 1 - \epsilon$ under $\mathscr{P}$,T and $\mathscr{Q}$,S respectively, and such that for any sequence in C_1 (or C_2), the ℓ sequences in it have distribution within $\frac{1}{10}\gamma$ of the distribution of ℓ sequences under $\mathscr{P}$,T (or $\mathscr{Q}$,S). If (as we assumed) $\bar{d}_2((\mathscr{P},T),(\mathscr{Q},S))$ were not $\geq \alpha - \gamma$, then ϵ could be chosen so that for infinitely many n there are a sequence in C_1^n and a sequence in C_2^n that differ in $< (\alpha - \gamma)n$ places. This implies that the distribution of ℓ sequences in the above two n sequences is within $\alpha - \gamma$ in the $\bar{d}$ metric, and hence the distribution of ℓ sequences under $\mathscr{P}$,T and $\mathscr{Q}$,S differs by $< \alpha - \frac{1}{2}\gamma$. The above would hold for all ℓ giving a contradiction. On the other hand, it is clear that $\bar{d}_2((\mathscr{P},T),(\mathscr{Q},T))$ cannot be $> \alpha + \gamma$ for $\gamma > 0$, because for each n, $\bar{d}(\{T^i\mathscr{P}\}_1^n\ \{S^i\mathscr{Q}\}_1^n) \leq \alpha$, and after removing two collections of sequences of measure ε, $\varepsilon < \frac{1}{2}$, $\varepsilon\alpha < \frac{1}{4}\gamma$ (in the Z-space), there are still sequences that match to within $(\alpha + \gamma)n$ terms.

REFERENCES

1. P. R. Halmos, "Lectures on ergodic theory," Publ. Math. Soc. Japan, No. 3, Math. Soc. Japan, Tokyo, 1956; reprint, Chelsea, New York, 1960, 54–55. MR 20 #3958.
2. P. Billingsley, Ergodic Theory and Information, Wiley, New York, 1965. MR 33 #254.
3. V. I. Arnold and A. Avez, Ergodic Problems of Classical Mechanics, Benjamin, New York, 1968. MR 38 #1233.
4. Jacobs, Lecture Notes on Ergodic Theory, vol. 1 and 2, Univ. of Aarhus, Denmark, 1962–63.
5. M. Smorodinsky, Ergodic Theory, Entropy. Springer Lecture Notes 214 (1970).
6. P. Shields, The Theory of Bernoulli Shifts, University of Chicago Press, Chicago and London, 1973.
7. D. S. Ornstein, "Bernoulli shifts with the same entropy are isomorphic," Advances in Math. 4 (1970), 337–352. MR 41 #1973.
8. _____, "Two Bernoulli shifts with infinite entropy are isomorphic," Advances in Math. 5 (1970), 339–348.
9. _____, "Factors of Bernoulli shifts are Bernoulli shifts," Advances in Math. 5 (1970), 349–364.
10. _____, "Imbedding Bernoulli shifts in flows," Contributions to Ergodic Theory and Probability, Lecture Notes in Math., Springer-Verlag, Berlin, 1970, 178–218.
11. _____, "The isomorphism theorem for Bernoulli flows," Advances in Math., March, (1973).
12. _____, "An example of a Kolmogorov automorphism that is not a Bernoulli shift," Advances in Math., (to appear).
13. _____, "A K-automorphism with no square root and Pinsker's conjecture," (to appear).
14. _____, "A mixing transformation for which Pinsker's conjecture fails," (to appear).
15. N. A. Friedman and D. S. Ornstein, "On isomorphism of weak Bernoulli transformations," Advances in Math. 5 (1970), 365–394.
16. P. Shields and D. S. Ornstein, "An uncountable family of K-automorphisms," Advances in Math., (to appear).
17. B. Weiss and D. S. Ornstein, "Geodesic flows are Bernoullian," Israel Journal of Math., (to appear).
18. M. Smorodinsky, "An exposition of Ornstein's isomorphism theorem," Advances in Math., (to appear).
19. W. Ambrose, "Representation of ergodic flows," Ann. of Math. 42 (1942), 723–739.

20. R. Adler and P. Shields, "Skew products of Bernoulli shifts with rotations," Israel Journal of Math., (to appear).
21. R. Adler, P. Shields, and M. Smorodinsky, "Irreducible Markov shifts," Ann. Math. Statist. 43 (1972), 1027–1029.
22. J. Feldman and M. Smorodinsky, "Bernoulli flows with infinite entropy," Ann. Math. Statist. 42 (1971), 381–382.
23. Y. Katznelson and B. Weiss, "Commuting measure preserving transformations," Israel Journal of Math., (to appear).
24. R. McCabe and P. Shields, "A Class of Markov shifts which are Bernoulli shifts," Advances in Math. 6 (1971), 323–328.
25. G. Gallavotti, "Ising model and Bernoulli schemes in one dimension," (preprint).
26. G. Maruyama, "Some aspects of Ornstein's theory of isomorphism problems in ergodic theory," (to appear).
27. H. Totoki, "On a class of special flows," presented at Symp. on Ergodic Theory, Mathematisches Forschungsinstitut Oberwolfach, 1968; Universitätsdruckerei H. Stürtz AG Würzburg, printed in Germany.
28. S. Ito, H. Murata, and H. Totoki, "Remarks on the isomorphism theorems for weak Bernoulli transformations in general case," reprinted from Publ. of the Res. Inst. for Math. Sciences, Kyoto Univ., vol. 7, No. 3, 1972.
29. S. M. Rudolfer and K. M. Wilkinson, "A number-theoretic class of weak Bernoulli transformations," (to appear).
30. K. M. Wilkinson, "Ergodic properties of certain linear mod one transformations," (to appear).
31. M. Smorodinsky, "β-Automorphisms are Bernoulli shifts," (to appear).
32. Ja. G. Sinai, "On the foundations of the ergodic hypothesis for a dynamical system of statistical mechanics," Dokl. Akad. Nauk SSSR 153 (1963), 1261–1264 = Soviet Math. Dokl. 4 (1963), 1818–1822. MR 35 #5576.
33. _____, "Dynamical systems with elastic reflections," Uspekhi Mat. Nauk. 27 (1972), 137f.
34. A. G. Kushnirenko, "Upper bound of the entropy of a classical dynamical system," Dokl. Akad. Nauk SSSR, 161 (1):37–38 (1965).
35. S. A. Juzvinskiĭ, "Metric properties of endomorphisms of compact groups," Izv. Akad. Nauk SSSR Ser. Mat. 29 (1965), 1295–1328; English transl., Amer. Math., Soc. Transl. (2) 66 (1968), 63–98. MR 33 #2798.
36. A. N. Kolmogorov, "A new metric invariant of transient dynamical systems and automorphisms in Lebesgue spaces," Dokl. Akad. Nauk SSSR 119 (1958), 861–864. (Russian) MR 21 #2035a.
37. _____, "Entropy per unit time as a metric invariant of automorphisms," Dokl. Akad. Nauk SSSR 124 (1959), 754–755. (Russian) MR 21 #2035b; 21, 1599.
38. Ja. G. Sinai, "A weak isomorphism of transformations with an invariant measure," Dokl. Akad. Nauk SSSR 147 (1962), 797–800 = Soviet Math. Dokl. 3 (1962), 1725–1729. MR 28 #5164a: 28 #1247.
39. L. D. Mešalkin, "A case of isomorphism of Bernoulli schemes," Dokl. Akad. Nauk SSSR 128 (1959), 41–44. (Russian) MR 22 #1650.
40. B. McMillan, "The basic theorems of information theory," Ann. Math. Statist. 24 (1958), 196–219.

41. J. Blum and D. Hanson, "On the isomorphism problem for Bernoulli schemes," Bull. Amer. Math. Soc. 69 (1963), 221–223.
42. C. Shannan, "A mathematical theory of communication," Bell System Tech. J. 27 (1948), 379–423, 623–656.
43. Ja. G. Sinai, "On the notion of entropy of a dynamical system," Dokl. Akad. Nauk SSSR 124 (1959), 768–771.
44. W. Krieger, "On entropy and generators of measure-preserving transformations," Trans. Amer. Math. Soc. 199 (1970), 453–464.
45. R. Adler and B. Weiss, "Similarity of automorphisms of the torus," Mem. Amer. Math. Soc. No. 98 (1970).
46. D. V. Anosov, "Geodesic Flows on Closed Riemannian Manifolds with Negative Curvature," Proc. of Steklov Inst. No. 90 (1967).
47. G. A. Hedlund, "The dynamics of geodesic flows," Bull. Amer. Math. Soc. 45 (1939), 241–260.
48. E. Hopf, "Statistik der Geodatischen Linien in Mannigfaltigkeiten negativer Krummung," Bertichte Sachs. Akad. Wiss. Leipzig 91 (1939), 261–304.
49. _____, "Statistik der Lösungen geodätischer Probleme vom unstabilen typus, II," Math. Ann. 117 (1940), 590–608.
50. Y. Katznelson, "Ergodic automorphism of T^n are Bernoulli shifts," Israel Journal of Math. 10 (1971), 186–195.
51. Pinsker, Dokl. Akad. Nauk, SSSR 133 (1960), 1025–1026.
52. V. A. Rokhlin, "Metric properties of endomorphisms of compact commutative groups," Izv. Akad. Nauk SSSR Ser. Mat. 28 (1964), 867–874.
53. V. A. Rokhlin and Ya. G. Sinai, "Construction and Properties of invariant measurable partitions," Dokl. Akad. Nauk, SSSR 141 (1961), 1038–1041.
54. J. Clark, "A Kolmogorov shift with no roots."
55. D. Ornstein and P. C. Shields, "Mixing Markov shifts of kernel type are Bernoulli," Advances in Math., (to appear).
56. I. M. Gel'fand and S. V. Fomin, "Geodesic flows on manifolds of constant negative curvature," Uspekhi Mat. Nauk, 49 (1):118–137 (1952).
57. Ya. G. Sinai, "Geodesic flows on compact surfaces of negative curvature," Dokl. Akad. Nauk SSSR, 136 (3):549–552 (1961).
58. Ya. G. Sinai, "Classical dynamical systems with countably multiple Lebesgue spectrum. II," Izv. Akad. Nauk SSSR Ser. Mat., 30 (10):15–68 (1966).
59. P. R. Halmos and H. Vaughn, "The marriage problem," Amer. J. Math. 72 (1950), 214–215.
60. W. Parry, Entropy and Generators in Ergodic Theory, Benjamin, New York, 1969.
61. V. A. Rokhlin, "Fundamental concepts of measure theory," Mat. Sb., 67 (10): 107–150 (1949).
62. Ornstein and Weiss, "Very Weak Bernoulli Implies Finitely Determined," (to appear).
63. _____, "Ergodic Theory of Z^n Actions with Applications to the Ising Model," (to appear).
64. Gallavotti and Ornstein, "The Billaird Flow with a Convex Scatterer is Bernoullian," (to appear).